The Economics of Managing
Crop Diversity On-farm

Issues in Agricultural Biodiversity

This series of books is published by Earthscan in association with Bioversity International. The aim of the series is to review the current state of knowledge in topical issues in agricultural biodiversity, to identify gaps in our knowledge base, to synthesize lessons learned and to propose future research and development actions. The overall objective is to increase the sustainable use of biodiversity in improving people's well-being and food and nutrition security. The series' scope is all aspects of agricultural biodiversity, ranging from conservation biology of genetic resources through social sciences to policy and legal aspects. It also covers the fields of research, education, communication and coordination, information management and knowledge sharing.

The Economics of Managing Crop Diversity On-farm

Case studies from the Genetic Resources Policy Initiative

Edited by
Edilegnaw Wale, Adam G. Drucker and Kerstin K. Zander

publishing for a sustainable future

London • Washington, DC

First published in 2011 by Earthscan

Earthscan Ltd, Dunstan House, 14a St Cross Street, London EC1N 8XA, UK

Earthscan LLC, 1616 P Street, NW, Washington, DC 20036, USA

Earthscan publishes in association with the International Institute for Environment and Development

For more information on Earthscan publications, see www.earthscan.co.uk or write to earthinfo@earthscan.co.uk

ISBN: 978-1-84971-221-7 hardback
ISBN: 978-1-84971-222-4 paperback

Typeset by MapSet Ltd, Gateshead, UK
Cover design by Adam Bohannon

A catalogue record for this book is available from the British Library

Library of Congress Cataloging-in-Publication Data

The economics of managing crop diversity on-farm : case studies from the Genetic Resources Policy Initiative / edited by Edilegnaw Wale, Adam G. Drucker and Kerstin K. Zander. — 1st ed.
     p. cm.
   Includes bibliographical references and index.
   ISBN 978-1-84971-221-7 (hardback) — ISBN 978-1-84971-222-4 (pbk.) 1. Crops—Germplasm resources—Economic aspects—Africa—Case studies. 2. Agrobiodiversity—Economic aspects—Africa—Case studies. 3. Plant varieties—Economic aspects—Africa—Case studies. 4. Genetic resources conservation—Economic aspects—Africa—Case studies. I. Edilegnaw Wale. II. Drucker, Adam G. III. Zander, Kerstin K.
   SB123.34.A35E26 2010
   338.1'62096—dc22

                                                                                              2010025027

Contents

4 Conclusions and Outlook

Contributors

Moses T. Daura is a lecturer and currently the head of Department of Animal Sciences in the School of Agricultural Sciences at University of Zambia. He earned his MSc in animal science at Kansas State University, USA (1980) and a PhD in animal nutrition at West Virginia University, USA (1989). His main research interests include environment and the well-being and performance of livestock, geopathic stress effect on livestock immune systems and the body's ability to efficiently utilize nutrients.
Contact: Moses T. Daura, Animal Sciences Department, University of Zambia, Lusaka, email: mtdaura@gmail.com

Adam G. Drucker, currently a senior (ecological) economist at Bioversity International, is interested in the economic implications of environmental impacts and the development of practical methodologies for the quantification of non-market values. He earned a PhD in environmental economics from Imperial College, London. He has been involved in a wide range of natural resource management issues including: sustainable agriculture and rural development, deforestation, genetic resource and biodiversity conservation (including livestock), water contamination, agrochemical use and human poisonings, climate change, environmental impact assessment and capacity building, participatory planning and appraisal, environmental policy and the use of market-based instruments. He has held a number of positions in Latin America, Africa and Australia with the United Nations, national governments, international NGOs, the University of London/UK Department for International Development, Charles Darwin University and the International Livestock Research Institute (ILRI, an institute of the Consultative Group on International Agricultural Research (CGIAR)).
Contact: Adam G. Drucker, Bioversity International, Maccarese, Rome, Italy, email: A.Drucker@cgiar.org

Jagadish Chandra Gautam is an agricultural economist with specialization in policy research focusing on rural development and management of natural resources. He is currently the executive chairman of Agriculture Business and Trade Promotion Multi-Purpose Cooperative (ABTRACO). He obtained a postgraduate diploma in agricultural economics (1971) and MSc in economics

(1973), University of New England, Australia. He served the government of Nepal for 35 years in various capacities such as agricultural planner and development officer.

Contact: Jagadish Chandra Gautam, Agri-Business and Trade Promotion Multipurpose Cooperative, 723/67 Tank Prasad Ghumtisadak, Kathmandu, Nepal, email: abtraco@hons.com.np

Sinafikeh Asrat Gemessa is currently a graduate student in public administration in international development at Harvard University's John F. Kennedy School of Government. He did his MSc in economics at Addis Ababa University (2008) and a BSc in statistics in the same university (2005). From May 2008 to December 2008, he worked as a research and office manager for the Agricultural Economics Society of Ethiopia (AESE). From January 2009 to June 2010, he was a research officer for the International Food Policy Research Institute (IFPRI) working mainly on agricultural productivity projects. His main research interests include production economics, microeconometrics and applying non-market valuation methods towards sustainable use and management of crop and animal genetic resources in Ethiopia in the developing world.

Contact: Sinafikeh Asrat Gemessa, Harvard University/John F. Kennedy School of Government, USA, email: sinafik12@yahoo.com

Elias Kuntashula is currently a PhD student in environmental economics at the Department of Agricultural Economics, University of Pretoria, South Africa. He worked as an on-farm research officer for the World Agroforestry Centre (ICRAF) for several years with a focus on socio-economic methods for analysis of agroforestry ecosystems on-farm. He has since 2005 been working as a lecturer/researcher at the University of Zambia Agricultural Economics Department. He did his MSc in agricultural economics at University of Zimbabwe in 1999. His main research interests include sustainable use and management of ecological systems through payments for environmental services.

Contact: Elias Kuntashula, Agri. Economics and Extension Education Dept, University of Zambia, Lusaka, email: ekuntashula@yahoo.com

Judith C. N. Lungu has been dean of the School of Agricultural Sciences at the University of Zambia since January 2005. She obtained a BSc at the University of Zambia (1975), MSc at University of Massachusetts, USA (1980) and PhD in animal physiology at University of Manitoba, Canada (1986). Before taking up the post as dean, she served as head of the Animal Sciences Department for six years. She has been lecturing in livestock sciences at University of Zambia since 1986. Her main research interests include food security, sustainable livestock systems and organic farming.

Contact: J.C.N. Lungu, Animal Sciences Department, University of Zambia, Lusaka, email: jlungu@unza.zm

Krishna Prasad Pant is presently working as a senior economist in the Ministry of Agriculture and Cooperatives, Kathmandu, Nepal. He has nearly two decades of experience in policy analysis, programme planning, research and teaching in agricultural and environmental economics. He completed his PhD in agricultural economics (1998) and worked as an economist and senior program officer for several years in the Ministry of Agriculture and Cooperatives, Nepal. He is teaching environmental economics in Kathmandu University as a Professor Adjunct where he supervised eight MSc students. He is associated with South Asian Network for Development and Environmental Economics (SANDEE) and also works with South Asian Network of Economic Institutions (SANEI). He is a member of International Association of Agricultural Economists (IAAE) and Nepal Agricultural Economics Society. He is presently working as a senior programme officer in the National Agriculture Research and Development Fund, Kathmandu. His main research interests include market solutions for climate change, valuation of agro-biodiversity and environmental services, environment and human health, biosecurity of agriculture and non-tariff barriers to food trade. *Contact:* Krishna Prasad Pant, Ministry of Agriculture and Cooperatives, Singha Durbar, Kathmandu, Nepal, email: kppant@yahoo.com

Edilegnaw Wale is currently a senior lecturer, School of Agriculture Sciences and Agribusiness, University of Kwa-Zulu Natal, South Africa. At the university, he teaches various courses and advises MSc and PhD students on their thesis work. Before joining the university in early 2009, he had been working with Genetic Resources Policy Initiative (GRPI) as an economist and has since been working closely with all GRPI partners on issues of economics as related to biodiversity. He holds a PhD in agricultural economics from the University of Bonn, Germany (2004) and MSc in agricultural development economics from Wageningen University, the Netherlands (1997). Having completed his BSc degree in agricultural economics (1993), he began his career at Alemaya University in Ethiopia where he served for six years in various capacities. He has also been a consultant to IFPRI and Bioversity (the then IPGRI), among others. Edilegnaw's major area of research interest is linking biodiversity with economic development at the micro level. He is a member of various professional associations and editor of journals including the *International Journal of Biodiversity and Conservation.*
Contact: Edilegnaw Wale, Department of Agricultural Economics, School of Agricultural Sciences and Agribusiness, University of KwaZulu-Natal, South Africa, email: walee@ukzn.ac.za

Kerstin K. Zander is a research fellow of the School for Environmental Research, Charles Darwin University, Australia. She did her MSc in agricultural science at the University of Bonn, Germany (2002) and a PhD in environmental and resource economics at the Center for Development Research (ZEF), University of Bonn (2006). Her main research interests include the economics of agro-biodiversity conservation, the economics of payments for environmental services and the economics of climate change adaptation with a focus on total economic values approaches.

Contact: Kerstin K. Zander, School for Environmental Research, Charles Darwin University, Darwin NT 0909, Australia, e-mail: kerstin.zander@cdu.edu.au

Preface

The purpose of this book is to document the range of economic issues of relevance to agro-biodiversity policy that have been identified in the various Genetic Resources Policy Initiative (GRPI) project countries. This work is the derived outcome of a participatory process during the implementation of that project. The partners and stakeholders engaged with the GRPI identified the issues under consideration and found the subsequent synthesis of the economics work useful for supporting their policy development processes.

The topics identified include trait and variety preferences (of farmers, consumers and traders); farmers' perceptions of the livelihood impacts of replacement and loss of traditional crop varieties; and the commercialization/marketing of, and value chain development for, traditional crop variety products. These are examined using empirical data from three of the GRPI project countries (Ethiopia, Nepal and Zambia) by applying a range of economic methods, which include choice experiments, hedonic pricing, contingent valuation and farm business income analysis.

The overall findings are relevant not only to the GRPI countries involved in the study but also to other countries concerned with the sustainable utilization of such resources. Given the importance of linking genetic resources conservation with farmers' incomes and survival strategies, this book will have achieved its objectives if it can illustrate how genetic resources issues can be integrated into development interventions to address potential policy trade-offs and, more importantly, if the issues addressed are picked up by the decision-makers in the respective GRPI countries.

Editors: Edilegnaw Wale, Adam G. Drucker and Kerstin K. Zander

List of Acronyms and Abbreviations

ABTRACO	Agriculture Business and Trade Promotion Multi-Purpose Cooperative
AESE	Agricultural Economics Society of Ethiopia
AnGR	animal genetic resource
CE	choice experiment
CGIAR	Consultative Group on International Agricultural Research
CL	conditional logit
CVM	contingent valuation method
DA	development agent
GRs	genetic resources
GRPI	Genetic Resources Policy Initiative
HPM	hedonic pricing method
IAAE	International Association of Agricultural Economists
ICRAF	International Centre for Research in Agroforestry (World Agroforestry Centre)
IFPRI	International Food Policy Research Institute
IIA	independence of irrelevant alternatives
ILRI	International Livestock Research Institute
IP	implicit price
MWTP	marginal willingness-to-pay
OLS	ordinary least squares
NARC	National Agricultural Research Council
PA	peasant association
PACS	payment for agro-biodiversity conservation services
PES	payment for environmental services
R&D	research and development
RF	rainfall
RPL	random parameter logit
SANDEE	South Asian Network for Development and Environmental Economics
SANEI	South Asian Network of Economic Institutions
TEV	total economic value
WTA	willingness-to-accept
WTP	willingness-to-pay
ZEF	Center for Development Research, University of Bonn

Part 1

Setting the Scene

Chapter 1

Introduction: Setting the Scene for GRPI Economics

Edilegnaw Wale

Background and rationale

Biological diversity (or biodiversity in short) is the number, variety and variability of all living organisms in terrestrial, marine and other aquatic ecosystems and the ecological complexes of which they are parts (UNCED, 1992). Conceptually, it encompasses both the number (stock and information contained therein) and variability dimensions (Wale, 2004). Agro-biodiversity is a subset of biodiversity relevant for agriculture and it covers the variability of plants, animals and micro-organisms. It can be considered at three levels, i.e. genetic, species and agro-ecosystems (Upreti and Upreti, 2002). It encompasses various biological resources tied to agriculture including edible plants and crops, livestock and fish, naturally occurring insects, bacteria and fungi, agro-ecosystem components, wild resources of natural habitats and landscapes, and the genetic resources contained therein (Thrupp, 2000). Crop diversity is a subset of agro-biodiversity relevant for crop production. All the above terms have been used in this book as relevant.

The agricultural sector depends on agro-biodiversity for sustainable agricultural production and the proper functioning of the agro-ecosystem. Agro-biodiversity has ecological, genetic, economic, scientific, educational and cultural values (Wale, 2004). Agro-biodiversity offers private (captured by farmers and consumers) and public (such as insurance[1] values, ecological services like resilience, etc.) benefits to society. Lipper and Cooper (2009) group the benefits of crop diversity into three main categories:

- Private benefits to farmers via the consumption and production values that they derive;

- Local or regional benefits to farmers and, ultimately, consumers when the choices make farming more resilient to biotic and a-biotic stress driven by agro-ecological changes;[2] and
- Global benefits to future farmers, plant breeders and consumers when the choices they make protect against genetic erosion.

Apart from these utilitarian arguments, there are also ethical and moral reasons to ensure the maintenance of biodiversity in general and agro-biodiversity in particular.

Narrow genetic stock implies higher susceptibility to any potential danger, especially for African agriculture which is dependent mainly on nature. Reduction in diversity often increases vulnerability to climatic and other stresses, raises risks for individual farmers and can undermine the stability of agriculture (Thrupp, 2000). For instance, if all farm households in a given area plant a single local variety of sorghum and if a certain disease occurs in that area which the variety cannot resist, then the negative socio-economic and agro-ecological consequences will be enormous. The loss of agro-biodiversity has inevitable risks and costs to agricultural productivity and food security (Thrupp, 2000), which matters everywhere (Perrings and Lovett, 1999). Therefore, conservation and sustainable use of agro-biodiversity will remain to be a key policy issue for the sustainable development of every country's agriculture.

Smallholder farmers of the developing world maintain the majority of the remaining agro-biodiversity. In developing countries, the lack of access to technologies has been the artificial cause of maintenance of traditional ways of farming by most smallholders (Hammer, 2003). However, for various reasons, conservation by farmers and the market is not enough (Wale, 2004). An individual farmer is not considering the number, quantity and identity of local varieties other farmers are maintaining on their farm. No farmer produces agro-biodiversity for its own sake and no farmer takes into account the effect of their actions on the level of agro-biodiversity regionally or nationally. Moreover, farmers' decisions are based on crop variety traits observable and relevant to them, which may not necessarily reflect all the public values discussed above. Farmers only produce crop diversity to the extent that it meets their private needs.

Due to differences between private (to farmers) and public (to society) values of crop genetic resources, the private optimum level of conservation will not be equal to the social optimum. It would be sheer coincidence if the level, composition and quality of diversity corresponded to the one demanded by society (Maier and Shobayashi, 2001). Farmers as a group underproduce crop diversity as a public good (Smale, 2006b) because they are not rewarded for their contribution to crop diversity with additional social (public) benefits (Kontoleon et al, 2009b). As noted above, crop diversity has both private and public benefit (quasi-public goods) dimensions and farmers will produce the aspect of diversity meeting their private demands. Thus, society cannot rely on farmers and market forces alone (Cooper et al, 2005c). Consequently, there is a conservation gap – a gap between what is maintained by the farmers/market and the level of agro-biodiversity that needs to be maintained to capture the three benefits listed above.

Though the causes, magnitude and consequences are context specific, there is now an increasing consensus that loss of crop genetic diversity is an ongoing rural development problem in developing countries. Genetic erosion is the single most serious strategic threat to the global food system (Gore, 1992). Habitat conversion (including the replacement of traditional varieties by improved/exotic ones), over-exploitation, trade and invasive species are the most important causes of biodiversity loss in general (Kontoleon et al, 2007a). Major causes of agro-biodiversity loss include degradation of agro-ecosystems, pollution, introduction of exotic species and genetic technologies, and selection pressure from human activities (Boef, 2000). Introduction of genetic technologies leads to the replacement of a large number of local varieties with a few, more uniform, high-yielding strains (Swaminathan, 2000) and deletion of indigenous species (Perrings and Lovett, 1999).

Publicly driven agro-biodiversity policy measures are, therefore, indispensable to address the conservation gap. In this book, agro-biodiversity policy refers to any course of action in which national regulations and guidelines are set for the operation of specific instruments aimed at sustainable use and conservation of agro-biodiversity and monitoring the extent to which those objectives are met. If a country is signatory to the international conventions/treaties (e.g. the International Treaty on Plant Genetic Resources for Food and Agriculture), it can include principles detailed in those treaties.

There are four main elements in a policy for biodiversity conservation (Perrings and Lovett, 1999):

- Regulatory regimes to protect key species, habitat and ecological services;
- An appropriate set of property rights in natural resources (along with their supporting institutions);
- A compensation mechanism; and
- A supporting structure of incentives and disincentives to induce the desired response.

Due to its scope, this volume will contribute mainly to the last two aspects. The first two are addressed in another Genetic Resources Policy Initiative (GRPI) book of this series.

Every country has, within its specific constraints and opportunities, different options to conserve these resources. The two main options can be classified into:

1 Public *ex situ* conservation strategies (cold room gene banks, botanical gardens, agricultural research stations and field gene banks); and
2 *In situ* conservation strategies in the form of farmers' *de facto* conservation,[3] conservation on farmers' fields through external incentives, national parks, and national reserves.

Of the different *in situ* conservation options, conservation on farmers' fields, also called on-farm conservation, has recently received considerable attention by governments, NGOs and the international community.

In situ conservation is gradually gaining greater acceptance because it is recognized that it is impractical to conserve all potentially useful genes *ex situ* for cost and technical reasons (Hawtin and Hodgkin, 1997). On-farm conservation, a subset of *in situ*, is also becoming more attractive. Its dynamic features, its capacity to maintain crop diversity and the indigenous knowledge associated with it, and the opportunity it opens up to link conservation and rural development are the typical desirable features of on-farm conservation. Poverty-ridden custodians of genetic wealth are increasingly confronted with severe socio-economic problems (Swaminathan, 2000). On-farm conservation offers a unique opportunity to link up conservation objectives with this poverty. Farmers participate in conservation initiatives if these activities support their livelihood strategies (Méndez et al, 2007). Agro-biodiversity conservation is not just a matter of ensuring the continuous survival of traditional varieties; it is a question of sustaining and enhancing the incomes and survival strategies of the rural people with which crop genetic resources are entwined. The challenge will be to develop cultivation systems that are a workable compromise between what is good for the farmers and what will benefit biological diversity (Perfecto et al, 1996).

Agro-biodiversity on farmers' fields is the outcome of utilization of indigenous varieties of crops on farmers' fields (Wale, 2008b). Since the use of traditional varieties on-farm automatically maintains genetic resources contained therein along with farmers' indigenous knowledge, the distinction between use and conservation is irrelevant for on-farm conservation. As noted above, in terms of the attention they deserve for conservation, one can identify two categories of traditional crop varieties: those with use value to farmers and hence are conserved *de facto*, and those which do not currently address farmers' concerns and are not maintained on-farm. Those elements of genetic resources of no current use value to farmers (but of public value for the future of agriculture) will have to be maintained through public conservation strategies. To these effects, there is a need to identify the kinds of varieties that can/cannot survive in the course of economic development interventions. Considering crop varieties as good as their attributes, the studies reported in the subsequent chapters (e.g. Chapters 2, 3 and 5) will generate information as to which types of variety traits are useful for farmers, consumers and traders. This information is meant to assist decision-makers to identify those varieties that will be maintained as the outcomes of agricultural production and those that need to be maintained via public conservation strategies.

This book has sought to document a variety of economic issues identified during the implementation of a global project – the Genetic Resources Policy Initiative. In Chapters 2–6 it presents the results of the various demand-driven topics identified in three of the six GRPI countries (Ethiopia, Nepal and Zambia). These empirical chapters are meant to highlight the mechanisms of linking genetic resources conservation, utilization and development.

Target audiences for this book include decision-makers and other stakehold-ers (mainly in GRPI countries) involved in the conservation and sustainable utilization of crop diversity. It is hoped and believed that agricultural/development economists and other professionals working on the management of agro-biodiver-sity can also benefit from this piece of work.

The rest of this introductory chapter is organized as follows. The ensuing section deals with the economics of genetic resources policy in some selected areas relevant to the book. This is followed by a discussion on the genesis of the GRPI-Economics work. The GRPI-Economics research/policy objectives and the research methodology are briefly presented subsequently. Finally, this chapter concludes presenting the road-map and overview of the empirical chapters.

Economics in genetic resources policy

Biodiversity economics refers to the economic analysis of the principles, causes and implications of changes in biological diversity (Kontoleon et al, 2009b). It deals with identifying the social benefits of agro-biodiversity conservation and the social opportunity costs that result from agro-biodiversity loss.

Economic analysis (e.g. research on the costs and benefits of maintaining agro-biodiversity) can feed into the multidisciplinary issue of agro-biodiversity policy. In general, the purpose of economic analysis (as it relates to agro-biodiver-sity) is to understand the linkages between policy interventions, autonomous changes and how all these changes influence farmers' preferences/incomes/decisions and agro-biodiversity outcomes. The critical element of the analysis is addressing the impact of economic development interventions in the conserva-tion, management and utilization of agro-biodiversity. To mitigate their impacts, the mechanisms through which development interventions affect farmers' crop-variety use decisions have to be understood and documented (Wale et al, 2009).

When a farmer chooses to adopt a new variety and replace an older variety, it reflects the farmer's judgment that the new variety offers some net benefit or advantage (Evenson and Gollin, 2003). To the more commercial and market-oriented farmers, abandoning production of local varieties may appear to be economically rational if returns from the improved varieties are higher. Other factors, such as subsidized inputs and services (e.g. improved varieties, fertilizer and extension advice) might also create incentives to grow exotic varieties. Subsidies on improved varieties, which do not reflect the social opportunity cost of the activity to which they apply, distort the comparative advantages and artifi-cially make traditional varieties less profitable to farmers (Pretty, 1995) while subverting farmers' expression of their own preferences and priorities (Chambers et al, 1989). Moreover, the profitability of modern agriculture is the outcome of free-riding on those farmers who are investing in such genetic diversity (Kontoleon et al, 2009b). Such policy distortions can ultimately undermine the farmers' contribution to maintaining traditional varieties of crops (Brown et al, 1993).

Economic analysis can help understand the incentives that farmers need in making the choice between raising local and/or improved varieties, as well as the identification of interventions compatible with the conservation and sustainable use of agro-biodiversity (Drucker and Anderson, 2004). Such analysis will inform policy-makers on the possible trade-offs (see 'Genetic resources policy trade-offs' below) between development and agro-biodiversity outcomes. In doing so, it helps to inform policy options that can achieve both agro-biodiversity conservation and poverty reduction.

Conceptual framework

The livelihoods literature (e.g. Ellis, 1998; Scoones, 1998; Bebbington, 1999) suggests that farmers respond to policy, environmental and resource constraints (e.g. by changing crop and variety choices). According to this literature, livelihoods diversification is the most widespread strategy. Livelihood diversification, which can be voluntary or involuntary, is defined as the process by which rural families construct a diverse portfolio of activities to survive and improve their standards of living (Ellis, 1998). Growing a portfolio of traditional varieties of crops is one of the most common strategies of farmers to address various concerns including risk.

The relationship between agricultural production and genetic resources is two directional, i.e. agricultural production affects genetic resources outcomes, and agricultural production and productivity also depend on the state of genetic resources (Day-Rubenstein and Heisey, 2001). Consider Figure 1.1 below which sketches the links among farmers' contextual characteristics, policy variables and crop diversity/seed technology use outcomes. It portrays the mechanisms through which farmers' decisions become genetic resources friendly or otherwise and the synergies/trade-offs involved. It also depicts the possible interlinkages between genetic resources policy-making and development interventions, issues addressed in the empirical case studies of the book.

According to Figure 1.1, farmers' resource endowments, institutions, policy and contextual characteristics influence their resource allocation behaviour. Endowments refer to the resources and rights that social actors have (Leach et al, 1999). Given their needs and constraints and the types of crop varieties (technologies) to which they have access, farmers follow certain livelihood strategies which involve resource allocation decisions. Seed technology choice, diversification and variety use are examples of these decisions. The end results are rural welfare (incomes, productivity, poverty, vulnerability, etc.) and natural resources outcomes, sustainable or unsustainable.

The impact of farmers' decisions on the outcome variables (natural resources and farmers' poverty/income conditions) reflects the trade-offs discussed in 'Genetic resources policy trade-offs' below. The synergy or conflict is context specific (Smale et al, 2006) as the conflicts between agriculture and biodiversity are by no means inevitable (Thrupp, 2000). The net effect of the trade-offs and synergies will result in genetic resources and human dimension outcomes which,

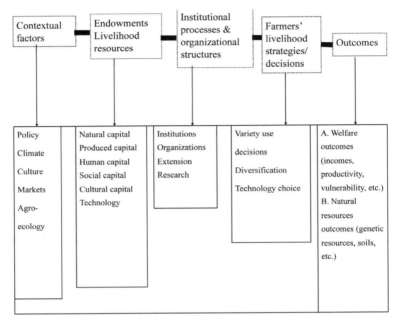

Figure 1.1 *Farmers' livelihoods and natural resources outcomes*

Source: Adapted from Scoones (1998) and Wale (2004)

in turn, will be affected by the contextual factors, endowments, institutions and farmers' livelihood strategies through the feedback loops and linkage effects.

Genetic resources valuation for policy

For the most part, policy-makers fail to understand the role of genetic resources for sustainable agricultural development. This could partly be because of the fact that the role/value of these resources in agriculture is not as visible as resources like soil and water. In other words, the loss in genetic resources and its impact on the national economy is invisible and less immediate but cumulative. To address this gap, valuation of agro-biodiversity conservation remains high on the policy agenda (Kontoleon et al, 2009b). In this book, valuation of genetic resources does not mean pricing in a cardinal sense but it is taken as a mechanism for the recognition of the diverse values of these resources in society and accounting for those values whenever possible (Wale, 2008a).

Valuation is meant to inform policy on the benefits of maintaining and the costs of losing genetic resources, and it is a means to justify investment on conservation. It is an input to undertake cost-benefit analysis of alternative policy options (Nunes and Nijkamp, 2008). In so doing, it entails addressing policy-relevant questions such as: Is conservation of biodiversity worth it? Does the minimum benefit outweigh the costs?

To the extent that the effort to value these resources does what the market fails to do and identifies the beneficiaries of these resources along the value chain, conservation agents can identify the agents of the loss (the would-be contributors to finance conservation) and the amount each agent has to contribute. Similarly, it also enables conservation agents to identify the agents of conservation who will have to be rewarded and decide on how much the reward should be. The valuation exercise is also an input to access and benefit-sharing legislations and bio-prospecting agreements so that the arrangements to share the benefits and implement the bio-prospecting agreements can be made in accordance with the value of the genetic material. In sum, valuation is important for all facets of public decision-making that impact upon biodiversity resources (Kontoleon et al, 2007a).

Genetic resources policy trade-offs

Wale et al (2009) have noted the inevitability of policy impact trade-offs in genetic resources policy-making in the context of the situation prevailing in the GRPI countries. The relevance of the concept of policy impact trade-offs emanates from the interlinkage and interdependence of agro-biodiversity conservation and agricultural development (Day-Rubenstein and Heisey, 2001). The recent book edited by Kontoleon, Pascual and Smale (2009a) echoed the lively debate surrounding the trade-off between the dissemination of high-yielding modern varieties and the potential erosion of plant genetic diversity, as the former is often argued to have the potential to induce genetic uniformity.

Trade-offs in the current volume refers to the gain in the agricultural development objective and the loss in agro-biodiversity conservation, and vice versa, due to any agricultural policy. Even though there are possible synergies between agro-biodiversity and agricultural development, this book will focus on the trade-offs aiming to contribute towards policies directed at averting agro-biodiversity loss.

Trade-offs are often experienced through the impact of the development interventions on the comparative advantages of local/improved varieties. Extinction is related to the process of development itself (Drucker and Rodriguez, 2009). In the absence of mitigating policy measures, negative externalities of development policies/projects (meant to take care of rural poverty through crop productivity) will be revealed through crop diversity loss. A lot of crop varieties have already been discarded as a result of policy incentives and consequently farmers' decisions to improve their livelihoods (de Ponti, 2004). For instance, adoption of wheat genetic technologies is resulting in wheat diversity loss in the central highlands of Ethiopia (Yifru and Hammer, 2006). Moreover, the loss in crop diversity (of food crops like sorghum) is sometimes due to displacement of the crop and its traditional varieties by other more rewarding cash crops such as khat (*Catha edulis*) in Ethiopia (Wale, 2004).[4]

On-farm conservation involves opportunity costs. These costs often increase with better access to markets, inputs and productive farming systems (Wale, 2008b). There is an inverse relationship between on-farm conservation outcomes and opportunity costs (Cooper et al, 2005b). In other words, less-favoured areas, which maintain the dominant stock of agro-biodiversity, are the ones that face lesser opportunity costs of growing diverse traditional crops and varieties. However, this does not imply that poor farmers will have to remain detached from markets and technological options to maintain traditional varieties. It rather shows the policy trade-offs decision-makers face.

Markets have major impacts on agricultural biodiversity, by affecting farmers' choice of crops and varieties to grow (Lipper et al, 2009). Market development often reduces the chance of cultivating traditional varieties (Smale, 2006a). Agricultural market linkage has often been associated with a number of negative impacts, including the reduction in on-farm crop genetic diversity and narrowing the crop genetic resource base of agricultural systems (Bellon, 2004). The specific trade-offs identified in different GRPI countries have been discussed in more detail in Wale et al (2009).

Despite all such trade-offs, policy-makers and donors do not often consider biodiversity impacts during the design of policies, projects and programmes (Wale et al, 2009). Development interventions often fail to internalize their unintended impacts on biodiversity (Srivastava et al, 1996). This problem is by and large the outcome of lack of understanding of the above interdependencies and linkages. Trade-offs between conservation of biodiversity and poverty reduction (food security) should be taken into account when designing conservation policies (Kitti et al, 2009). To deal with trade-offs, decision-makers will have to consider and internalize genetic resources issues when they formulate other development policies and programmes. They also have to place mitigating measures when development policies are in conflict with genetic resources. Otherwise, agricultural development interventions that erode crop genetic diversity will become self-defeating in the long-run as crop genetic resources are inputs for genetic technology development. The case studies presented in this book analyse some of the contextual trade-offs and suggest possible mechanisms to arrive at the 'right but hard' mix of policies, choices and decisions.

Value chains and value addition as a strategy for genetic resources policy

Value chains[5] can be defined as the sequence of value-adding activities leading to end use (Sturgeon, 2001). It encompasses all activities involving input supply, production, processing and marketing. Due to its holistic nature, no problem will be left untouched and this makes it a suitable avenue to address the preceding trade-offs in a manner that adds value to the products of traditional crop varieties so that farmers can benefit from these varieties and diversity outcomes can be improved.

Why and how do research and development (R&D) in value chains address the conservation of agro-biodiversity? Policies to promote *in situ* conservation include promotion of demand for products of diverse (landrace) varieties (Cooper et al, 2005b). One of the innovative elements of R&D on value chains for crop genetic resources is the opportunity it offers to link farmers' incomes, their survival strategies and the diversity of those crops.[6] Value addition to the products of traditional varieties of crops will create extra market demand for the crop as an intermediate product. This will, in turn, increase its market price and improve farmers' income, increase their incentive to grow the traditional crop varieties and ensure the sustainable cultivation of the crop. Instead of directly compensating farmers the financial opportunity costs, linking the products of farmers' varieties to agricultural value chains is more attractive on sustainability and feasibility grounds (Wale, 2008b).

Addressing the institutional and transaction cost problem along the value chain can enhance the cultivation, processing and marketing of those traditional varieties with a value-adding role in the chain. However, not all landraces are equally valuable (Smale, 2006b). Market value chain analysis (e.g. Chapter 6 in this volume) can inform policy decisions as to which types of traditional varieties will be more useful along the value chain. As it could be that all varieties of a given crop are not suitable for value addition, value chains may discriminate against certain varieties, i.e. the use and diversity of some varieties can be enhanced at the expense of the others. This calls for further R&D on this subject, which is becoming increasingly important.

The genesis of GRPI-Economics work

The problem of losing genetic resources in developing countries, among other things, can be attributed to policy and institutional failures. By and large, biodiversity loss is a problem of inadequate institutions and incentives (Kontoleon et al, 2007a). The failures have occurred because the policies do not lead to the desirable outcomes set in advance, or rural development policies do not account for the agro-biodiversity impacts, or the recommendations are not taken up altogether.

In any attempt to formulate and implement biodiversity policy, technical, institutional and capacity constraints are prevalent, especially in developing countries where most of the world's genetic resources are found. Taking these imperatives into account, the GRPI project, among other things, was initiated to strengthen capacity in genetic resources policy in six selected developing countries (Egypt, Ethiopia, Nepal, Peru, Vietnam and Zambia) and three regions (East Africa, West and Central Africa, and the Andean regions). To contribute to capacity-building in the economics of genetic resources policy issues, a postdoctoral researcher and a graduate student (MSc) were supported by the project to do data collection and empirical analysis in one of the GRPI countries, Ethiopia.

GRPI, among other things, was designed to assist decision-makers with measures that can reduce negative impacts of development policy measures on genetic resources. It was designed to contribute towards cost-effective, stakeholder-sensitive and community-based genetic resources conservation and utilization strategies. The GRPI-Economics work, as part of this international initiative, initiated in Phase 2 of the project, aimed to feed economic analysis results into genetic resources policy-making.

The economics research in the GRPI project has been underway in Ethiopia, Nepal and Zambia since June 2005. Taking into account the 3M (multidisciplinary, multi-stakeholder and multi-sectoral)[7] nature of genetic resources policy, it was found critical to engage the relevant stakeholders.

Although a number of studies and books have been produced on the economics of agro-biodiversity conservation (e.g. Smale, 2006a; Cooper et al, 2005a; Kontoleon et al, 2007b; and Kontoleon et al, 2009a), there is little evidence of economic analysis being 'mainstreamed' in the policy-making process. Relative to other research areas of public policy, economics has contributed little to debates about the value of these resources (Smale, 2006b). Experiences in terms of integrating economic analysis results into the policy process are very scarce. Previous studies do not appear to have had an impact on policy proposals in different countries.

One of the tenets of this current volume is that, among other things, this is due to the lack of participatory process that engages all the relevant stakeholders in the identification and examination of the problem(s). It is recognized during the implementation of GRPI project that it is imperative for decision-makers to understand how economic policies and programmes in different sectors affect agro-biodiversity conservation outcomes.

Hoping to make the recommendations more appealing to decision-makers, GRPI launched a participatory process during the early stages; the specific issues/topics addressed are identified by stakeholders themselves (decision-makers, government organizations, researchers from various disciplines, NGOs and grass-root organizations, farmers and private businesses) in the respective GRPI countries. Engaging various stakeholders in the respective countries, some of which have conflicting mandates and vested interests, has been entrusted to enhance the willingness and ability of different stakeholders to take collective actions so as to improve both farmers' incomes and agro-biodiversity outcomes.

The case studies reported in this book are not topics identified by the authors of the respective chapters. Like many other GRPI products, this book is the outcome of a demand-driven process during the implementation of GRPI. It is, therefore, expected that the results and their implications will be integrated into the policy processes in the respective countries. Consequently, the GRPI team firmly believes that the products will not be shelved but taken up by the decision-makers in the respective GRPI countries. That is how this volume and other GRPI products in this series aim to elevate the profile and agro-biodiversity policy impact of research in various fields (economics, institutional analysis, legal analysis, etc.).

GRPI-Economics research and policy objectives

The general objective of GRPI-Economics is to conduct economic analysis on policy issues identified through the process outlined in the section above so that the results can feed into harmonizing genetic resources policy trade-offs. To this end, as the subset of the global multidisciplinary project, GRPI-Economics had the following specific demand-driven objectives:

- to study farmers' preferences/choices to identify crop varieties maintained *de facto* and those that need policy interventions;
- to examine the values (mainly trait-based) of traditional varieties of crops;
- to explore farmers' views and perceptions on replacement (of traditional varieties of crops by improved ones) and loss, and the importance of the outcome of this process to their livelihoods;
- to examine the potentials (to genetic resources policy) of value addition and commercialization of traditional varieties of crops; and
- to generate policy relevant information that will contribute towards developing a policy framework for the value addition, commercialization and incentive design for the sustainable utilization and conservation of traditional varieties of crops.

GRPI-Economics research methodology

Data collection

Household survey data were collected in 2007 in some of the GRPI countries (Ethiopia, Nepal and Zambia) to address the economics research questions identified in the respective countries. In Ethiopia and Nepal, multi-stage stratified random sampling (to identify regions, districts, villages, local communities and farm households studied in the respective countries) was applied. In Zambia, there was no formal sampling as consumers/traders of maize and groundnut were identified more purposively in various areas of urban Lusaka. It has to be re-emphasized that all the case studies in this volume have implications limited to areas with similar contextual features.

Data were collected through interviews with individual farmers/consumers/traders using questionnaires (different in content in each country). The questionnaires were pre-tested in all countries. Moreover, key informant interviews (with elderly farmers, extension agents, community leaders, local officials, women's groups and members of local informal institutions) and group discussions were held as necessary. Secondary data were also gathered and consultations made with policy-makers, development practitioners and agricultural researchers in each country.

Methods of data analysis

Broadly speaking, drawing from the environmental economics literature, there are two main sets of methods (stated and revealed) for addressing genetic resources policy issues to which economic analysis can contribute. In stated preference methods, individuals (economic agents) are given options to state their preferences by making choices from a range of defined alternatives which represent hypothetical market situations/scenarios. Choice experiments (as in Chapter 2), exploratory studies that deal with views, perceptions and attitudes (as in Chapter 4),[8] and variety attribute preference ranking (as in Chapter 5) fall under stated preference methods. The revealed preference methods, on the other hand, explore individuals' decision behaviour in the real world by observing their transactions in markets. In this class of methods, one can think of approaches such as hedonic pricing (as in Chapter 3) and analysis of outcomes – gross margin analysis, value chain analysis and farm business income analysis of alternative choices (for instance, in Chapter 6).

While choice experiments have frequently been applied in animal genetic resources literature (e.g. Scarpa et al, 2003a, 2003b; Tano et al, 2003, Zander and Drucker, 2008), they are not that often applied in the crop diversity literature. Chapter 2 is one of the first studies of its kind to apply choice experiments to crop diversity. In this chapter, choice experiments were designed to elicit farmers' preferences for a range of discrete variety attributes (yield, price, yield stability and environmental adaptability) identified from the information gathered from the rapid rural appraisal and key informant interviews. Every effort has been made to make the choices simple and clear for farmers to compare. For instance, instead of offering the households with high-yielding and marketable varieties, precise yield levels and market prices were set to elicit their preferences. This has created the opportunity to understand the trade-offs that farmers make between non-monetary variety attributes (yield stability and early maturity) and money convertible traits (yield).

Chapter 3 evaluates different useful traits of rice landraces in Nepal using attribute preference analysis. A hedonic pricing model is employed disaggregating the prices paid by the consumers for different useful traits of rice. The model captures consumers' willingness-to-pay for each rice variety trait. For estimating farmers' derived demand for seeds with different useful traits, a contingent valuation method is employed. This model captures farmers' bid for seeds of new varieties of rice with desirable traits. Chapter 4 analyses farmers' perceptions and views on loss and replacement of traditional varieties of crops in Ethiopia using descriptive statistics and a simple logit analysis. The estimated model is used to identify characteristics that distinguish farmers of different perceptions on issues of loss and what it means to their respective households.

In the Zambian case study, the maize and groundnut data are analysed using variety attribute preference ranking and regression analysis. The attribute rates were subjected to paired t-test analysis while regression analysis was conducted to understand the influence of different contextual characteristics of households on the quantity of local maize (Gankata) purchased per week.

Chapter 6 utilizes farm business income as the measure of the financial viability of rice landraces to farmers. Farm business income is computed as the income to the farm family from the crop after deducting the cost of purchased inputs from the total gross income.

The valuation concept taken in this volume is largely the one followed in Smale (2006a): the use of economic concepts to explain and predict human choices. If one defines valuation more broadly as the use of economic concepts to explain the role of traditional varieties of crops in agricultural R&D and predict the impact of human choices on the future of traditional varieties of crops, all the chapters can be considered as valuation case studies.

The road map and overview of the chapters

This small volume is organized around seven chapters and four parts. The five empirical chapters are grouped according to the methodological approaches adopted, the subject matters addressed and the policy implications drawn. The book will present the economics of managing crop diversity on-farm in terms of crop variety attribute preferences, value addition and marketing of the products of traditional crop varieties.

The absence of traits preferred by farmers and the presence of unacceptable traits are variety-dependent qualities that determine farmers' use of crop varieties. Identification of varieties maintained *de facto* and those which need policy interventions is done in all case study countries: Zambia, Nepal and Ethiopia. The diversity of traditional rice varieties is valued in Nepal and the commercialization potentials of traditional varieties are examined in Nepal and Zambia.

Chapter 1 has so far set the scene by introducing the subject matter of the book. It has explained the need for agro-biodiversity policy and the rationale for the volume. The genesis of the book, the objectives and the methodology are presented subsequently.

In Part 2, the book turns to variety trait preferences, perceptions and on-farm conservation policy. Chapter 2 of this part analyses Ethiopian farmers' preferences for sorghum and teff variety traits. It begins establishing the relationship between farmers' concerns and variety attribute preferences. This is followed by the theoretical underpinnings behind the choice experiment approach. Having described the study sites and sampled farm households, the findings are explained in the second half of the chapter. The analysis has generated policy-relevant information on the relative importance of crop-variety attributes to farm households. It has also made it possible to capture yield and price levels that farmers are willing to sacrifice to get non-monetary traits like yield stability and environmental adaptability. This further sheds light on how non-monetary variety attributes can be converted into their monetary values.

In Chapter 3, having discussed the approaches for measuring the economic value of crop diversity, the authors estimate the economic value of rice landraces. This case study documents the different useful traits of rice landraces in Nepal. Accordingly, Nepalese farmers attach a substantial value to a combination of high-yielding and aromatic traits, enough to justify investment in conservation.

Chapter 4 investigates farmers' views on the problems of replacement and loss of traditional varieties. It also looks into whether or not farmers view replacement and loss relevant to their livelihoods. This chapter starts presenting the scientific controversies on the questions of replacement and loss mainly drawing from the literature. Having established the research gap – integrating farmers' views and perceptions into this discussion – it then discusses the data collection process and analyses the data elicited from 395 farmers. Based on the empirical results, the chapter concludes that future on-farm conservation strategies have to target farmers based on their perceptions of replacement and loss, and the importance of the loss to their livelihoods.

Part 3 then deals with market value chains, commercialization and on-farm conservation policy. This part has two chapters from two GRPI countries. Chapter 5 deals with consumers' attribute preferences and traders' challenges affecting the use of local maize and groundnut varieties in Zambia. Recognizing the links among on-farm utilization of traditional crop production and consumption, this case study examines preferences of consumers and traders. This is done drawing from trader and consumer surveys on both maize and groundnut. The results have revealed that before making decisions to buy maize, consumers pay more attention to quantity-related attributes (such as grain size and kernel density) than to quality-related attributes (such as food taste). In contrast, when consumers make decisions to buy groundnut, quality-attributes were perceived as important. It is concluded that, when dealing with staple foods such as maize, breeders should focus on quantitative traits whereas qualitative traits should be targeted for non-staple food crops such as groundnut.

Chapter 6 deals with value addition and commercialization of rice landraces in Nepal. The results of the input cost and income analyses are then presented subsequently. This chapter explains farmers' rationale for engaging themselves in growing traditional crop varieties despite net loss if all the inputs, including home-produced resources, were to be imputed and costed. Based on the empirical results, the chapter suggests possibilities for promoting value-addition enterprises for rice and some selected local varieties of crops.

Part 4 concludes the book and offers outlook on the possible ways forward for economics and genetic resources policy. It recaps the issues, summarizes the major findings and draws the implications of the results for genetic resources policy.

Notes

1   Conservation of a diverse portfolio of traditional crop varieties safeguards agriculture from potential losses (due to disease outbreaks, pathogens, pests, and vagaries of climate) that could arise in the foreseeable future.

2   The loss of every gene and species limits our options for the future to breed varieties responsive to changes in climate, for instance (Swaminathan, 1996). The need to adapt to climate change and increased variability of production conditions is likely to increase farm-level demands for crop diversity (Lipper et al, 2009).

3   *De facto* conservation is the decision of farmers to continue cultivating local varieties (Meng, 1997) without any external public inducement. It is, in short, the positive externality of farming which, as noted above, will not result in the optimum level of crop genetic resources.

4   This is an economically important stimulant cash crop in many parts of Ethiopia, especially southeast. Consumers chew the leaves.
5   Value chains are also called commodity chains, activity chains, production chains or supply chains.
6   While there are R&D activities on the poverty and growth dimensions of agricultural value chains, there is hardly any study that links agricultural value chains with agro-biodiversity outcomes.
7   For details on the 3M methodology, see Wale et al (2009).
8   Such studies can also generate information on farmers' views/perceptions about a certain genetic resource policy question that permits farmers' perspectives to be integrated into crop genetic resources policy issues.

References

Bebbington, A. (1999) 'Capitals and capabilities: a framework for analysing peasant viability, rural livelihoods and poverty', *World Development*, vol 27, no 12, pp2021–2044

Bellon, M. (2004) 'Conceptualizing interventions to support on-farm genetic resource conservation', *World Development*, vol 32, no 1, pp159–172

Boef, W. S. (2000) 'Learning about institutional frameworks that support farmer management of agro-biodiversity: tales from the unpredictables', PhD thesis, Wageningen University, Wageningen, Netherlands

Brown, K., Pearce, D., Perrings, C. and Swanson, T. (1993) 'Economics and the conservation of global diversity', Working Paper Number 2, Global Environmental Facility, Washington, DC

Chambers, R., Pacey, A. and Thrupp, L. (eds) (1989) *Farmer First: Farmer Innovation and Agricultural Research*, ITDG, London

Cooper, J. C., Lipper, L. M. and Zilberman, D. (eds) (2005a) *Agricultural Biodiversity and Biotechnology in Economic Development*, Springer, New York

Cooper, J. C., Lipper, L. M. and Zilberman, D. (2005b) 'Introduction, agricultural biodiversity and biotechnology: economic issues and framework for analysis', in J. C. Cooper, L. M. Lipper and D. Zilberman (eds) *Agricultural Biodiversity and Biotechnology in Economic Development*, Springer, New York, pp3–13

Cooper, J. C., Lipper, L. M. and Zilberman, D. (2005c) 'Synthesis chapter: managing plant genetic diversity and agricultural biotechnology for development', in J. C. Cooper, L. M. Lipper and D. Zilberman (eds) *Agricultural Biodiversity and Biotechnology in Economic Development*, Springer, New York, pp457–477

Day-Rubenstein, K. and Heisey, P. (2001) 'Agricultural resources and environmental indicators: crop genetic resources', *ERS Agricultural Resources and Environmental Indicators*, Agricultural Handbook no AH722, Chapter 3.2

de Ponti, T. (2004) 'Combining on-farm PGR conservation and rural development: clash or synergy?', MSc thesis, Wageningen University, Wageningen, Netherlands

Drucker, A. and Anderson, S. (2004) 'Economic analysis of animal genetic resources and the use of rural appraisal methods: lessons from Southeast Mexico', *International Journal of Agricultural Sustainability*, vol 2, no 2, pp77–97

Drucker, A. G. and Rodriguez, L. C. (2009) 'Development, intensification and the conservation and use of farm animal genetic resources' in A. Kontoleon, U. Pascual and M. Smale (eds) *Agro-biodiversity Conservation and Economic Development*, Routledge, London and New York, pp92–110

Ellis, F. (1998) 'Household strategies and rural livelihood diversification', *Journal of Development Studies*, vol 35, no 1, pp1–38

Evenson, R. E. and Gollin, D. (2003) 'Assessing the impact of the green revolution, 1960–2000', *Science*, vol 300, no 5620, pp758–762

Gore, A. (1992) *Earth in the Balance: Ecology and the Human Spirit*, Houghton Mifflin, Boston, MA

Hammer, K. (2003) 'A paradigm shift in the discipline of plant genetic resources', *Genetic Resources and Crop Evolution*, no 50, pp3–10

Hawtin, G. C. and Hodgkin, T. (1997) 'Towards the future', in N. Maxted, B. V. Ford-Lloyd and J. G. Hawkes (eds) *Plant Genetic Conservation: The In Situ Approach*, Chapman and Hall, London, pp368–383

Kitti, M., Heikkila, J. and Huhtala, A. (2009) 'Agro-biodiversity in poor countries: price premiums deemed to miss multifaceted targets?', in A. Kontoleon, U. Pascual and M. Smale (eds) *Agro-biodiversity Conservation and Economic Development*, Routledge, London and New York, pp293–317

Kontoleon, A., Pascual, U. and Smale, M. (eds) (2009a), *Agro-biodiversity Conservation and Economic Development*, Routledge, London and New York

Kontoleon, A., Pascual, U. and Smale, M. (2009b) 'Agrobiodiversity for economic development: what do we know?', in A. Kontoleon, U. Pascual and M. Smale (eds) *Agro-biodiversity Conservation and Economic Development*, Routledge, London and New York, pp1–24

Kontoleon, A., Pascual, U. and Swanson, T. (2007a) 'Introduction', in A. Kontoleon, U. Pascual and T. Swanson (eds) *Biodiversity Economics: Principles, Methods and Applications*, Cambridge: Cambridge University Press, pp1–21

Kontoleon, A., Pascual, U. and Swanson, T. (2007b) *Biodiversity Economics: Principles, Methods and Applications*, Cambridge: Cambridge University Press

Leach, M., Merns, R. and Scoones, I. (1999) 'Environmental entitlements: dynamics and institutions in community-based natural resource management', *World Development*, vol 27, no 2, pp225–247

Lipper, L. and Cooper, D. (2009) 'Managing plant genetic resources for sustainable use in food and agriculture: balancing the benefits in the field', in A. Kontoleon, U. Pascual and M. Smale (eds) *Agro-biodiversity Conservation and Economic Development*, Routledge, London and New York, NY pp27–39

Lipper, L., Dalton, T. J., Anderson, C. L. and Keleman, A. (2009) 'Agricultural markets and sustainable utilization of crop genetic resources', in L. Lipper, C. L. Anderson and T. J. Dalton (eds) *Seed Trade in Rural Markets: Implications for Crop Diversity and Agricultural Development*, Earthscan, London, and Food and Agriculture Organization of the United Nations, Rome, pp1–20

Maier, L. and Shobayashi, M. (2001) *Multifunctionality: Towards an Analytical Framework*, Organization for Economic Co-operation and Development, Paris

Méndez, V. E., Giessman, S. R. and Gilbert, G. S. (2007) 'Tree biodiversity in farmer cooperatives of a shade coffee landscape in western El Salvador', *Agriculture, Ecosystems and Environment*, no 119, pp145–159

Meng, E. C. H. (1997) 'Land allocation and in situ conservation of crop genetic resources: the case of wheat landraces in Turkey', PhD dissertation, University of California, Davis, CA

Nunes, P. A. L. D. and Nijkamp, P. (2008) 'Introduction to the special issue on biodiversity policy', *Ecological Economics*, no 67, pp159–161

Perfecto, I., Rice, R. A., Greenberg, R. and van der Voort, M. E. (1996) 'Shade coffee: a disappearing refuge for biodiversity', *Bioscience*, vol 46, no 8, pp598–608

Perrings, C. and Lovett, J. (1999) 'Policies for biodiversity conservation: the case of Sub-Saharan Africa', *International Affairs*, vol 75, no 2, pp281–305

Pretty, J. (1995) *Regenerating Agriculture: Policies and Practice for Sustainability and Self-reliance*, Earthscan, London

Scarpa, R., Drucker, A. G., Anderson, S., Ferraes-Ehuan, N., Gomez, V., Risopatron, C. and Rubio-Leonel, O. (2003a) 'Valuing animal genetic resources in peasant economies: the case of the Box Keken creole pig in Yucatan', *Ecological Economics*, vol 45, no 3, pp427–443

Scarpa, R., Kristjanson, P., Ruto, E., Radeny, M., Drucker, A. G. and Rege, J. E. O. (2003b) 'Valuing indigenous farm animal genetic resources in Africa: A comparison of stated and revealed preference estimates', *Ecological Economics*, vol 45, no 3, pp409–426

Scoones, I. (1998) 'Sustainable rural livelihoods: a framework for analysis', Working Paper 72, Institute for Development Studies, Brighton, UK

Smale, M. (ed.) (2006a) *Valuing Crop Biodiversity: On-farm Genetic Resources and Economic Change*, CABI Publishing, Wallingford, UK

Smale, M. (2006b) 'Concepts, metrics and plan of the book', in M. Smale (ed.) *Valuing Crop Biodiversity: On-farm Genetic Resources and Economic Change*, CABI Publishing, Wallingford, UK, pp1–16

Smale, M., Lipper, L. and Koundouri, P. (2006) 'Scope, limitations and future directions' in M. Smale (ed.) *Valuing Crop Biodiversity: On-farm Genetic Resources and Economic Change*, CABI Publishing, Wallingford, UK, pp280–295

Srivastava, J., Smith, N. J. H. and Forno, D. (1996) *Biodiversity and agriculture: implications for conservation and development*, World Bank Technical Paper no 321, World Bank, Washington, DC

Sturgeon, T. J. (2001) 'How do we define value chains and production networks?', www.inti.gov.ar/cadenasdevalor/sturgeon.pdf, accessed 20 November 2008

Swaminathan, M. (1996) 'Compensating farmers and communities through a global fund for biodiversity conservation for sustainable food security', *Diversity*, no 12, pp73–75

Swaminathan, M. S. (2000) 'Government-industry-civil society: partnerships in integrated gene management', *Ambio*, vol 29, no 2, pp115–121

Tano, K., Faminow, M., Kamuanga, M. and Swallow, B. (2003) 'Using conjoint analysis to estimate farmers' preferences for cattle traits in West Africa', *Ecological Economics*, vol 45, no 3, pp393–407

Thrupp, L. A. (2000) 'Linking agricultural biodiversity and food security: the valuable role of agro-biodiversity for sustainable agriculture', *International Affairs*, vol 76, no 2, pp265–281

UNCED (1992) *Convention on Biological Diversity*, United Nations Conference on Environment and Development, Geneva.

Upreti, B. R. and Upreti, Y. G. (2002) 'Factors leading to agro-biodiversity loss in developing countries: the case of Nepal', *Biodiversity and Conservation*, no 11, pp1607–1621

Wale, E. (2004) 'The economics of on-farm conservation of crop diversity in Ethiopia: incentives, attribute preferences, and opportunity costs of maintaining local varieties of crops', PhD thesis, University of Bonn, Bonn, Germany

Wale, E. (2008a) 'Challenges in genetic resources policy making: some lessons from participatory policy research with a special reference to Ethiopia', *Biodiversity and Conservation*, vol 17, no 1, pp21–33

Wale, E. (2008b) 'A study on financial opportunity costs of growing local varieties of sorghum in Ethiopia: implications for on-farm conservation policy', *Ecological Economics*, vol 64, no 3, pp603–610

Wale, E., Chishakwe, N. and Lewis-Lettington, R. (2009) 'Cultivating participatory policy processes for genetic resources policy: lessons from the Genetic Resources Policy Initiative (GRPI) project', *Biodiversity and Conservation*, vol 18, no 1, pp 1–18

Yifru, T. and Hammer, K. (2006) 'Farmers' perception and genetic erosion of tetraploid wheat landraces in Ethiopia', *Generic Resources and Crop Evolution*, no 53, pp1099–1113

Zander, K. K. and Drucker, A. G. (2008) 'Conserving what's important: using choice model scenarios to value local cattle breeds in East Africa', *Ecological Economics*, no 68, pp34–45

Part 2

Variety Trait Preferences and On-farm Conservation Policy

Chapter 2

Economic Analysis of Ethiopian Farmers' Preferences for Crop Variety Attributes: A Choice Experiment Approach

Sinafikeh Asrat Gemessa and Kerstin K. Zander

Summary

Ethiopia has immense wealth of crop genetic resources, which is due to its diverse agro-ecology and cultural diversity. The country's genetic resources are, however, subject to serious erosion and irreversible losses due to policy, institutional and market failures. This study aims to contribute to a better understanding of the challenges with bearings on the sustainable management of crop genetic diversity through analysing farmers' crop variety attribute preferences and identifying the key socio-economic factors that condition their attribute preferences. The study applied a choice experiment (CE) method to elicit preferences and estimate the relative importance of the attributes in defining the perceived utility to be derived from four traits of sorghum and teff varieties, the two major food crops in the country (as a source of staple food for many parts of the country, teff, an annual grass, is primarily grown to prepare Ethiopian bread known as 'injera', porridge and some native alcoholic drinks). The attributes included selling price, productivity, environmental adaptability (resistance to drought, poor soil and frost occurrences) and yield stability of the variety despite occurrences of disease and pest problems. The analysis of farmers' preferences was based on primary data collected from farmers growing 131 teff and sorghum in the northeastern part of Ethiopia. Farm households attached the highest private value to the environmental adaptability trait of both sorghum and teff crops. This was followed by yield stability and productivity attributes of the same crops. The results also reveal that differences between farm households, in terms of household characteristics, their endowments and constraints, and the level of development integration (in the areas of basic infrastructure and agricultural extension) affect farmers' private valuation of crop variety attributes. Based on the empirical results, the chapter derives policy implications in the areas of on-farm conservation and improved variety use in Ethiopia.

Introduction

Many societies around the world depend on agricultural innovation processes for their daily livelihoods, particularly food supply. Crop genetic resources, embodied in the diversity of seeds planted by farmers, are important to secure food production and supply (di Falco et al, 2010). Different stakeholders such as farmers and breeders as well as gene-bank managers and crop scientists draw on diverse crop genetic resources to innovate, support and benefit society as a whole (Smale, 2006).

Sustainable management of crop genetic resources means assuring their diversity, both in trust collections or gene banks (*ex situ*) and on farms (*in situ*) (Smale, 2006; Bezabih, 2008). For farmers crop biodiversity is important to combat risks, from plant diseases to pests, and to adapt to changing production systems and changing environments (di Falco et al, 2010). Crop biodiversity is essential for the functioning of ecosystems and the provision of ecosystem services (e.g. Naeem et al, 1994) and also provides dietary needs and services that consumers demand as economies change (Smale, 2006).

Crop genetic resources are environmental goods that are renewable but vulnerable to losses from either natural or human-made interventions. Crop genetic improvement and the increase in farm inputs, such as pesticides and fertilizers, were driven by the goal to increase yield and yield stability, and have transformed rural societies in many parts of the world (Smale, 2006). There is, however, a growing concern about potential loss of crop biodiversity associated with social and economic change. The common challenge now is to develop strategies that enable crop genetic resources to be managed in ways that satisfy the needs of farmers and consumers at present and in the future. Crop biodiversity is a quasi-public good (Bezabih, 2008) and its conservation also benefits society as a whole, including future generations. The provision of crop biodiversity is largely on-farm, in particular in low-input agricultural systems in developing countries, and therefore the level of conservation is highly dependent on the preferences and decisions of farmers.

The purpose of this study is to contribute to a better understanding of the challenges of crop genetic conservation by providing an insight into Ethiopian farmers' preferences for crop variety attributes and to identify the key socio-economic characteristics that influence these preferences. Ethiopian policy-makers have to be informed about, *inter alia*, 'Who prefers what kind of variety attributes the most?' and 'How much are farmers willing to trade-off one variety attribute for another?' if on-farm conservation programmes are to be undertaken successfully. This study deals with these two questions by analysing farmers' attribute preferences, expressed by their willingness to pay for varieties of the two major crops in the country: sorghum and teff.

Similar questions about the WTP for certain attributes and about preference heterogeneity have already been raised in the context of evaluating animal genetic resources (AnGRs). These studies have attracted considerable attention, starting off with the work by Scarpa et al (2003a, 2003b). The most recent studies in East Africa are on the evaluation of cattle (Zander and Drucker, 2008), goats (Omondi et al, 2008a) and sheep (Omondi et al, 2008b), all based on choice experiments. For crop genetic resources, we are aware of only a few studies that use CE. Birol et al (2006) applied a CE to study farmers' demand for agricultural biodiversity in the home gardens of Hungary's transition economy, while Ruto and Garrod (2009) investigated farmers' preferences in a wider perspective, not on crop attributes but on agri-environment schemes. It is, therefore, important to conduct further research using stated preference methods such as CE on crops and their preferred attributes. Having knowledge about the holistic value of Ethiopian indigenous flora and fauna, and in particular its total agro-biodiversity, can inform conservation priority setting. Understanding the role of farmers in cultivating specific local crops or keeping unique farm animals in the conservation process is essential if improved varieties/breeds developed by breeding programmes are to address the concerns of farmers and their well-being.

For this study, a CE was conducted to estimate the individual utility farmers derive from four attributes of sorghum and teff varieties including:

- producers' price;
- productivity;
- environmental adaptability; and
- yield stability.

In a CE, individuals are given a hypothetical setting and asked to choose their preferred alternative among several options in a choice set, and they are usually asked to perform a sequence of such choices. Each alternative (a teff or sorghum variety in this case) was described by a number of attributes and their levels. Crop varieties possess private as well as public benefits, together accounting for the total economic value of crop genetic resources. Maintaining local crop varieties in Ethiopia provides a quasi-public good benefit with the external effect of conserving a genetic pool that has global significance for breeding and biodiversity. However, given the approach taken in this study, only internal (private) values to farmers, which are mainly use values, were captured. Assessing such use values plays a key role in orienting conservation and breeding strategies as conventional economic analyses often ignore the importance of indirect use values (e.g. socio-cultural/medical use, their ability to withstand biotic and abiotic stresses) associated with local varieties (see Zander and Drucker, 2008). The empirical analysis of farmers' preferences for the above attributes was based on primary data collected from farmers growing 131 teff and sorghum in the northeastern part of Ethiopia.

The remaining part of this paper is structured as follows. A relationship between farmers' concerns and variety attribute preferences is drawn in the next section. The third section outlines the theoretical underpinnings behind the choice experiment approach. The fourth section explains the data generation process followed by a description of the study sites and sampled farm households. The design and administration of the choice experiment is explained in the fifth section. The penultimate section discusses the findings from the analysis of the choice experiment data. Policy implications are drawn in the final section.

Farmers' concerns and preferences for variety attributes

Understanding farmers' preferences for crop attributes and their incentives to grow diverse varieties are critical to the success of on-farm conservation (di Falco et al, 2010). Such an understanding will also help in the areas of research priority-setting and improved breeding (Wale and Yalew, 2007). Preferences for variety attributes are, in turn, shaped by farmers' economic (resource constraints, markets and risk) and non-economic (religion, culture, norms and attitudes) concerns. For example, when local variety attributes satisfy farmers' concerns, their *de facto* conservation is the outcome of the correspondence between variety attributes and farmers' concerns. The preference for a variety and land allocation decision is dependent on farm household characteristics, their attitudes and concerns (Wale, 2004).

The probability that farmers choose to cultivate a certain crop variety, on the other hand, is dependent on the key attributes the farmers associate with the variety. Each farmer, however, derives different utility from consuming different varieties with different attributes based on the farm household's characteristics and attitudes. The survival of a variety on-farm, or the successful adoption of a newly introduced improved variety, is mainly a function of the maximized benefit to the farm household (Wale and Yalew, 2007).

The choice experiment and welfare measures

Since most of the attributes that characterize the varieties of crops are not directly tradable, non-market valuation methods must be used to determine their relative economic value. These benefits primarily accrue to farmers in non-market values or utility. The preferences of farmers, who are both producers and consumers of crop variety outputs, determine the implicit values they attach to crop varieties and their attributes (Louviere et al, 2000).

Of the range of environmental valuation approaches, the choice experiment method is appropriate for valuing crop varieties, considering their multiple benefits and functions. This method makes possible the estimation not only of the value of the environmental asset as a whole, but also of the implicit values of its attributes (Hanley et al, 1998; Bateman et al, 2003).

The approach, upon which the framework for choice modelling is based, has a theoretical grounding in Lancaster's model of consumer choice (Lancaster, 1966), and an econometric basis in models of random utility (Luce, 1959; McFadden, 1974). From random utility models, welfare measures can be obtained, expressed as farmers' willingness-to-pay (WTP) or willingness-to-accept (WTA) compensation for a change in crop varieties' attribute levels. The estimates for these welfare measurements are obtained from applying a conditional logit (CL) model, whose specification is detailed in many textbooks (e.g. Greene, 2000; Freeman, 2003). With one attribute being price, the implicit price (IP) for a change in any attribute, all else being equal, can be calculated. If the IP for an attribute is negative, then we obtain a WTA estimate, because farmers will be worse-off with a utility change and require compensation to be left at the same utility level. If the IP has a positive sign, then farmers have a WTP for the attribute in question. The IP is calculated by the ratio of coefficients of the attributes in question $\beta_{attribute}$, as obtained from the CL model and the coefficient of the monetary variable $\beta_{monetaryvariable}$ (see e.g. Rolfe et al, 2000; Zander and Holm-Mueller, 2007).

$$IP = -\frac{\beta_{attribute}}{\beta_{monetaryvariable}}$$

(Eqn 2.1)

The assumptions about the distributions-of-error terms implicit in the use of the conditional logit model impose the independence of irrelevant alternatives (IIA) property (Louviere et al, 2000). This property states that the probability of a particular alternative being chosen is independent of other alternatives, whether or not IIA property holds can be tested by dropping an alternative from the choice set and comparing parameter vectors for significant differences (Louviere et al, 2000). A common test to detect violation of the IIA property is the Hausman test (Hausman and McFadden, 1984), as applied to our data.

The data generation process

Survey data was drawn from farmers residing in two peasant associations (PAs)[1] in the northeastern part of Ethiopia (North Wollo zone of Amhara Regional State). Two phases of data collection procedures were implemented for this study within the framework of the Ethiopian component of Bioversity's GRPI project. As noted in Chapter 1, this project aimed to support the development of policy options for sustainable conservation and utilization of crop genetic resources in

Table 2.1 *Summary of main study site characteristics*

Study site characteristics	Woinye PA	Ala Wuha PA
Agro-ecological coverage	Midland 83%, Highland 10% and Lowland 7%	Midland 5% and Lowland 95%
Most important food crops	Teff, sorghum, finger millet, maize, wheat and barley	Teff, sorghum, maize and cow beans
Livestock assets owned by an average household in the PA	1 ox, 1 cow, 2 calves, 3 sheep and 3 goats	2 oxen, 2 cows, 2 calves and 4 goats

Source: Agricultural bureaus in Woinye and Ala Wuha PAs

Ethiopia. All the socio-economic characteristics employed in this study were collected in the first phase of data collection (from October 2006 to January 2007). Piloting of the first draft of the CE questionnaire and the actual CE survey were conducted in the second phase during June and July of 2007.

Stratified multistage sampling was adopted to identify Zones, Districts, PAs, villages and farm households. Overall, a total of 131 farmers were selected and interviewed from the two PAs found in the Guba Lafto district of North Wollo zone. The next part discusses the characteristics of the selected study sites (PAs) covered in this study. See 'Design and administration of the choice experiment' section below for the design of the CE survey.

Study site description

A summary of the main characteristics of the two PAs surveyed is reported in Table 2.1. Teff, sorghum and maize were among the most important food crops in both PAs. Agro-ecologically, the midland area (locally known as 'Woina dega') is the dominant agro-ecology in Woinye PA, covering 83 per cent, whereas the lowland area (locally known as 'Kola') is the major agro-ecology in Ala Wuha PA, covering 95 per cent. This should, however, increase the representativeness of our surveyed farm households since our sample covers farmers from the three major agro-ecologies of the country (midland, highland and lowland) growing the two major crops (sorghum and teff).

Farm household characteristics in North Wollo

The characteristics of the surveyed households and farm decision-makers are indicated in Table 2.2. The descriptive statistics for the binary variables (e.g. Gender) are reported in percentage terms. Assuming that the variables reported in Table 2.2 have the same direction of influence on preferences of attributes of both crops, their hypothesized effects on the demand for attributes considered in this study are also included in Table 2.2. Definition of each farm household characteristic reported in Table 2.2 is given below:

Table 2.2 *Descriptive statistics of farm household contextual characteristics and their hypothesized effects on the demand for attributes of crop varieties*

Characteristics	Mean (SD) N = 131	Producers' price	Productivity	Environmental adaptability	Yield stability
Household characteristics					
Gender	90.1%	±	±	±	±
Household size	5.38 (2.04)	+	+	+	+
Experience	25.38 (11.64)	+	+	+	+
Off-farm work	32.3%	+	+	–	–
No. dependants	1.15 (1.45)	+	+	+	+
Poverty status	85.5%	+	+	–	–
Farm and livestock characteristics					
Land shortage	64.8%	+	+	+	+
Total land size	0.75 (0.52)	±	±	±	±
Livestock value	5016.5 (4745.5)	+	+	–	–
Development integration Characteristics					
Access services	48.24 (27.07)	–	–	+	+
Agri. extension	70.2%	+	+	–	–

Source: GRPI, Ethiopian survey, 2006/2007

1 gender of the household head (denoted as 'Sex' in the model estimation, where 1 denotes male and 0 denotes female);

2 the number of household members who share the same food stock (denoted as 'Household size');

3 farming experience of the household head in years (denoted as 'Experience');

4 whether or not any member of the farm household works off-farm (denoted as 'Off-farm work', where 1 denotes at least one member working off-farm and 0 otherwise);

5 whether or not the farm household has been participating in the agricultural extension package programme (denoted as 'Agri. extension', where 1 denotes participating and 0 otherwise);

6 average walking distance (in minutes) the household head takes to reach electricity, piped water, telephone, primary school, secondary school, all-weather roads and irrigation infrastructures (denoted as 'Access services');[2]

7 whether or not the household head considers land shortage as the most important problem facing the household (denoted as 'Land shortage', where 1 denotes land shortage considered as the most important problem and 0 otherwise);

8 total land size operated by the household in hectares (denoted as 'Total land size');

9 total value of livestock (including hives and poultry), in Birr,[3] that is currently owned by the household (denoted as 'Livestock value');

10 whether or not the household considers itself to be at least self-sufficient in
relation to other households in the area (denoted as 'Poverty status', where a
value of 1 means the households consider themselves to be self-sufficient and
0 if they consider themselves poor or very poor); and

11 number of dependants with no labour or money contribution in the household (denoted as 'No. dependants').

The average characteristics suggested that a typical farm household in North
Wollo zone was male-headed and medium-sized with six members, two of which
were economically dependent. The experience of the primary decision-maker was
about 25 years. The typical farm household had no members working off-farm,
lived 50 minutes walking distance away from basic infrastructures and partici-
pated in the agricultural extension programme. The average land size operated by
a farm household was 0.75 hectares and most farm households considered
scarcity of land as the primary problem. The average farm family had 5000 Birr
worth of livestock (including hives and poultry).

Choice experiment design and administration

Setting the scene: attributes and levels for the choice experiment

The crop variety attributes and levels used in this study are reported in Tables 2.3
and 2.4. These very important attributes and their levels were identified in consultation with experts (crop breeders and researchers with hands-on experience and
practical knowledge of the relevant variety attributes), by reviewing previous
studies and historical data, and by identifying the most important seed selection
criteria put forward by a focus group of surveyed farm households and extension
workers in the villages. Apart from their importance to farmers, these attributes
('Producers' price', 'Productivity', 'Environmental adaptability' and 'Yield stability') are also policy-relevant for designing an incentive mechanism to undertake
on-farm conservation ventures at least cost (for example, by identifying farmers
who are demanding attributes embedded in local varieties) or for successful rural
interventions like crop variety development and diffusion.

Inclusion of monetary attribute(s) is necessary for the welfare analysis (see
'The choice experiment and welfare measures' section above). Producers' price
and productivity attributes can be used as a direct monetary attribute or as a
proxy for monetary attribute depending on the socio-economic setup of farmers
participating in the choice experiment survey. For farmers actively participating
in the local markets by supplying their teff and/or sorghum output to the local
market, it would be appropriate to use producers' prices as direct monetary attribute. However, for farmers whose output is less than or just enough to satisfy their
household food consumption needs, productivity seems to be more appropriate
as a proxy for monetary attribute. The levels for these attributes were based on the

Zone's minimum, average and maximum values of producers' price and productivity of the crops during the last decade.

With more than 92 per cent of the surveyed households reporting that they have faced drought problems at least once during the last ten years, the choice of environmental adaptability trait of both crops was appropriate. The same can be said about the attribute yield stability of both crops: about 90 per cent of the surveyed households stated that they have faced disease or pest problems (causes of yield instability in our attribute definition) at least once during the last ten years.[4] These attributes had two levels representing the existence or absence of the attributes in each crop (see Tables 2.3 and 2.4).

Table 2.3 *Sorghum variety attributes and their levels used in the choice experiment*

Variety attributes	Definition	Attribute levels
Producers' price	The amount of money the farmer earns by selling 100kg of harvested sorghum of a particular sorghum variety	110 Birr, 150 Birr, 200 Birr
Productivity	Average production harvested per hectare from planting a particular sorghum variety	14 quintals/hectare, 19 quintals/hectare, 25 quintals/hectare
Environmental adaptability	Whether the variety is adaptable/tolerant to drought, poor soils and frost	The variety is adaptable vs the variety is not adaptable
Yield stability	Whether the variety gives stable yield year-after-year, despite occurrences of crop disease and pest problems, in the absence of drought and frost	The variety gives stable yield year-after-year vs the variety gives variable yield year-after-year

Table 2.4 *Teff variety attributes and their levels used in the choice experiment*

Variety attributes	Definition	Attribute levels
Producers' price	The amount of money the farmer earns by selling 100kg of harvested teff of a particular teff variety	210 Birr, 270 Birr, 330 Birr
Productivity	Average output harvested per hectare from planting a particular teff variety	8 quintals/hectare, 15 quintals/hectare, 20 quintals/hectare
Environmental adaptability	Whether the variety is adaptable/tolerant to drought, poor soils and frost	The variety is adaptable vs the variety is not adaptable
Yield stability	Whether the variety gives stable yield year-after-year, despite occurrences of crop disease and pest problems, in the absence of drought and frost	The variety gives stable yield year-after-year vs the variety gives variable yield year-after-year

Design and administration of the choice experiment

A large number of unique crop variety profiles can be constructed from these set of attributes and levels.[5] However, in this study, fractional factorial design was used to capture only the main effects, yielding nine alternatives which were allocated to different choice sets.[6] These nine alternatives were created using an orthogonal design.[7] The choice sets were then completed using a cyclical design principle (Bunch et al, 1996). A cyclical design is a straightforward extension of the orthogonal approach. First, each of the alternatives from a fractional factorial design is allocated to different choice sets. Attributes of the additional alternatives were then constructed by cyclically adding alternatives into the choice set based on the attribute levels. That is, the attribute level in the new alternative became the next, higher attribute level to the one applied in the previous alternative. If the highest level was attained, the attribute level was set to its lowest level (Carlsson et al, 2007).

We then assigned the initially created 9 alternatives from our fractional factorial design to nine choice sets and constructed 2 other alternatives per choice set (hence 18 others) following the procedure mentioned above. In total, we constructed 27 alternatives for sorghum and 27 alternatives for teff divided between 9 choice sets per crop. An example of a choice set is presented in Figure 2.1.

Sorghum Variety Characteristics	Sorghum Variety 1	Sorghum Variety 2	Sorghum Variety 3
Producers' price	150	200	110
Productivity	14	19	25
Environmentally Adaptable	Yes	No	Yes
Stable–in–yield	No	Yes	No

I prefer to plant **Sorghum variety 1**..... **Sorghum variety 2**.... **Sorghum variety 3** ......

(Please tick (√) one option)

Figure 2.1 *Sample choice set for sorghum*

To check the relevance of the choice experiment questions about local conditions, farmers' expectations and level of understanding, the questionnaires were pre-tested on a focus group of 16 farmers (8 from each PA). The pre-test results were discussed with the enumerators and necessary changes were made, taking into account farmers' responses. During the actual data collection, enumerators explained, using the local language, the context in which choices were to be made; that attributes of crop varieties had been selected as a result of prior research and were combined artificially; and defined each attribute and choice set using visual aids to ensure uniformity.. Respondents were informed that completion of the exercise would help agricultural policy-makers in the design of variety development and local variety conservation interventions. Out of the 131 households interviewed for the choice experiment survey, 66 of them were randomly chosen and presented with choice sets containing sorghum variety options while the remaining 65 were given teff variety options. All of the surveyed households answered all of the 9 choice sets (either sorghum or teff version) presented to them and hence a total of 1179 choices were elicited from our survey.

Bateman et al (2003) suggest restricting the number of attributes chosen for the design to a relatively small number (such as 4, 5 or 6). This is because the minimum required sample size increases exponentially in the number of attributes. Given our constraint to a relatively small sample size of about 130, we hence decided to include 4 attributes in the profiles.

Results and discussions

The choice experiment was designed with the assumption that the observable utility function would follow a strictly additive form (Louviere et al, 2000). The model was specified so that the probability of selecting a particular crop variety was a function of attributes of that variety. That is, for the population represented by the sample, indirect utility from crop variety attributes takes the form of Equation 2.2:

$$V_{ij} = \beta_0 + \beta_1 Z_{pprice} + \beta_2 Z_{productivity} + \beta_3 Z_{adaptability} + \beta_4 Z_{yield-stability}$$

(Eqn 2.2)

where β_{1-4} refer to the vector of coefficients associated with the vector of attributes describing crop variety attributes and β_0 is the alternative specific constant.

To begin with the estimation of Equation 2.2, two conditional logit models were fitted for each crop (for either teff or sorghum variety options). The Independence of Irrelevant Alternative (IIA) property was tested, which is implicit in the error structure of the conditional logit (CL) model, using the Hausman and McFadden (1984) test contained within LIMDEP Nlogit. The tests, however, provided inconclusive results for both crops by failing to find a positive definite difference matrix for any two alternatives; and this was the case for all three tests conducted by dropping a different alternative each time,

Table 2.5 *Random parameter logit estimates for choice of variety,*
with standard errors (SE) in parentheses

Variable	Sorghum		Teff	
	Coeff. (SE)	Coeff. Stdv.	Coeff. (SE)	Coeff. Stdv.
Alternative 1	0.364** (0.163)	–	0.613*** (0.150)	–
Alternative 2	1.293*** (0.271)	–	0.887*** (0.263)	–
Producers' price	1.841*** (0.225)	–	0.862*** (0.149)	–
Productivity	0.272*** (0.024)	–	0.217*** (0.018)	–
Environmental adaptability	4.703***	2.920***	4.446***	3.290***
	(0.720)	(0.606)	(0.718)	(0.701)
Yield stability	4.220***	2.6257***	3.1060***	2.654***
	(0.660)	(0.583)	(0.617)	(0.587)
Number of observations	594		585	
ρ^2	0.566		0.530	
Log likelihood	−283.263		−301.915	

Notes: *** significant at 1% level; ** significant at 5% level; * significant at 10% level.
Source: GRPI, Ethiopian survey, 2006/2007

indicating that the models do not fully conform to the underlying IIA property. Models that relax the IIA property, such as Random Parameter Logit model (RPL, also referred to as Mixed Logit), have to be estimated as it is done in this paper (Hensher et al, 2005). In the RPL model estimated for each crop, all of the attributes except for the monetary attribute (producers' price) and the proxy for monetary attribute (productivity) were defined to be normally distributed. The models were estimated with simulated maximum likelihood with Halton draws using 500 replications (see Train, 2003 for details on simulated maximum likelihood and Halton draws). The models were estimated using Nlogit 4.0.

Although the experiment was generic, we included two alternative specific constants, since the purpose was to test if there were any other factors than the attributes themselves that affected farmers' variety choices. The results are presented in Table 2.5.

The results in Table 2.5 show that all of the sorghum and teff variety attributes were highly statistically significant factors in the choice of both crop varieties, and that they had the expected signs in that the fulfilment of any single attribute increases the probability that a sorghum (or teff) variety was selected, other attributes remaining equal. The overall fit of the model for each crop, measured by McFadden's ρ^2, was good. The estimated standard deviations of the random parameters were also significant, and in relation to the mean estimates they were not as large as the mean coefficients, suggesting relatively low preferences heterogeneity for these attributes. The only unexpected finding was that the two alternative specific constants were significant. This indicates that all else being equal, respondents were more likely to choose alternative 1 or 2, compared to alternative 3. This might be due to the design of the choice experiment questions.

Table 2.6 *Mean MWTP for each variety attribute by crop and type of monetary attribute (standard errors in parentheses)*

Attribute	$MWTP_1$		$MWTP_2$	
	Sorghum	Teff	Sorghum	Teff
Productivity	14.77	25.16	–	–
	(1.756)	(4.40)		
Environmental adaptability	255.50	515.66	17.29	20.50
	(42.501)	(111.851)	(2.557)	(3.324)
Yield stability	229.27	360.28	15.52	14.32
	(38.773)	(88.905)	(2.320)	(2.830)

Notes: $MWTP_1$: marginal willingness-to-pay values measured in terms of Birr per quintal of the respective crop (producers' price used as the monetary attribute). $MWTP_2$: marginal willingness-to-pay values measured in terms of quintals of the respective crop per hectare (productivity attribute used as a proxy for the monetary attribute).
Source: GRPI, Ethiopian survey, 2006/2007

In Table 2.6 we report the estimated mean marginal willingness-to-pay (MWTP) for each of the attributes. These are simply the ratio between the attribute coefficient and producers' price coefficient (expressed by $MWTP_1$) or the ratio between the attribute coefficient and coefficient for productivity (expressed by $MWTP_2$). Note that the attributes for environmental adaptability and yield stability were binary variables, and thus they could be directly compared. For productivity, it is the MWTP in Birr for an increase in productivity by 1 quintal per hectare.

The productivity attribute may also be used as a proxy for monetary attribute, and may even be more appropriate in cases where only a small portion, if any, of the agricultural output of a farm family makes it to the market after satisfying household food consumption needs of the family. The $MWTP_2$ values reported in Table 2.6 are based on productivity attribute taken as a proxy for monetary attribute.

The results of both measures of MWTP showed that farm households in North Wollo zone seemed to be very risk averse since they were willing to pay a rather substantial amount for more adaptable and/or stable varieties of both crops. This is perhaps reflected in their strong willingness to diversify the crops they plant between different kinds of traditional and improved varieties to buffer the impact of drought and/or disease problems.

The $MWTP_1$ and $MWTP_2$ values for environmental adaptability were higher than their counterparts for yield stability for both crops, and for teff the difference in WTP was significant using a t-test. The $MWTP_1$ values for the productivity attribute showed that respondents were are willing to pay 15 Birr and 25 Birr for an increase in productivity by 1 quintal per hectare.

To account for observed heterogeneity of preferences across farm households, we also estimated models where a set of socio-economic characteristics are interacted with the attributes. However, in random utility models the effects of social and economic characteristics on choice cannot be examined in isolation but as interaction terms with choice attributes. Due to possible multicollinearity problems, it was not possible to include all the interactions between the explanatory variables collected in our survey and the four crop variety attributes when estimating the random logit models with interactions (Breffle and Morey, 2000). The results of the two models with socio-economic characteristics are presented in Tables 2.7 and 2.8.

The results in Table 2.7 showed that the interaction between the demand for higher levels of productivity in sorghum varieties and the sex of the household head was positive. This showed that male-headed households demanded more productive sorghum varieties than female-headed households. This may be because households with male heads have larger sizes (and hence demand more output from their land) than households with female heads and those females usually assume this position in a family when they are either widowed or separated from their husbands.[8]

Farm households with at least one member working off-farm demanded more productive sorghum varieties compared to those households with no members working off-farm. At least two explanations can be forwarded here. First, the opportunity cost of labour in crop production is higher for farm households with an off-farm job opportunity compared to those without, reflected in their higher demand for highly productive sorghum varieties. Second, production of sorghum by resource-poor farmers is usually at least partly for home consumption. However, the percentage of sorghum grain produced and then marketed may be greater for farm households with off-farm job opportunity since they are more likely to be better integrated into the local markets, prompting them to demand higher productivity from their sorghum variety options to get more marketable surplus. The results in Tables 2.7 and 2.8 also showed that farm households with more experienced heads demanded higher environmental adaptability trait from both sorghum and teff variety options. In the drought-prone areas of North Wollo zone (such as the PAs covered in this survey), more experienced farmers were likely to have gone through a greater number of recurrent drought encounters in the past, inducing them to look for varieties that are better resistant to such environmental pressures.

The results in Table 2.7 may also shed light on why farmers choose to participate in the agricultural extension package programme, with the positive interaction term between productivity attribute and agricultural extension participation. Farmers might be motivated to participate in the extension because they demand high-yielding sorghum varieties from these services.

The results in Table 2.7 also showed that farmers operating a relatively large land size also demanded less environmental adaptability trait in sorghum varieties compared to those operating smaller lands. Smaller land size can be translated into smaller total output and less scope to diversify into different crop varieties, and farmers were particularly risk averse towards non-adaptable varieties planted on these plots since, otherwise, they put at risk the much needed output that these plots provide to the vagaries of nature.

Table 2.7 *Random parameter logit estimates for sorghum variety traits interacted with socio-economic characteristics*

Variable	Coefficient	St. error	Coeff. Stdv	St. error
Random parameters				
Yield	−0.097	0.126	0.117***	0.035
Environmental adaptability	7.847	354.403	2.065***	0.606
Yield stability	11.898	354.405	2.370***	0.590
Non-random parameters				
Alternative 1	0.188	0.196		
Producers' price	0.019***	0.003		
Alternative 2	1.691***	0.332		
Heterogeneity in mean parameters				
Productivity* Sex	0.146*	0.082		
Productivity* Off-farm work	0.113*	0.068		
Productivity* Agri. extension	0.111*	0.065		
Env. adaptability* Experience	0.137**	0.065		
Env. adaptability* Land size	−3.519***	1.325		
Number of observations	*513*			
ρ^2	*0.611*			
Log likelihood	*−219.231*			

Notes: *** significant at 1% level; ** significant at 5% level; * significant at 10% level.
Source: GRPI, Ethiopian survey, 2006/2007

The results of the RPL model for teff variety choices with socio-economic characteristics are presented in Table 2.8.

The RPL results for teff variety choices showed that farmers with larger land size to operate also demanded more productive teff varieties compared to those operating smaller lands. This is unexpected because with more than 63 per cent of the surveyed households reporting land shortage as a primary problem, households with smaller land sizes were expected to compensate for this by demanding more productive teff varieties. This might be because teff is a highly commercial crop and the perceived utility from more productive teff varieties was higher for farm households operating larger land sizes and who were likely to produce a greater proportion of their output for the market, i.e. marketable surplus.

The results further showed that farmers who reported higher drought frequency in the past also demanded more productive teff varieties compared to those with fewer drought encounters. This might reflect their uncertainty about the future production prospect and the need to hoard maximum teff production output for household consumption in the coming season.

Households with larger livestock assets demanded less environmentally adaptable and stable yielding teff varieties compared to those with smaller livestock assets. Crop production is the single most important source of livelihood for farmers who cannot rely on their livestock assets as an insurance against crop

Table 2.8 *Random parameter logit estimates for teff variety traits interacted with socio-economic characteristics*

Variable	Coefficient	St. error	Coeff. Stdv	St. error
Random parameters				
Yield	0.160	0.207	0.190***	0.038
Environmental adaptability	−10.252*	5.411	2.950***	0.980
Yield stability	7.787	6.124	3.826***	1.264
Non-random parameters				
Alternative 1	0.513***	0.186		
Producers' price	0.012***	0.002		
Alternative 2	1.379***	0.344		
Heterogeneity in mean parameters				
Productivity* Land size	0.276***	0.095		
Productivity* Drought frequency	0.088***	0.030		
Env. adaptability* Livestock value	−0.553***	0.178		
Env. adaptability* Household size	1.440**	0.657		
Env. adaptability* Experience	0.096*	0.052		
Yield stability* Livestock value	−0.396**	0.163		
Number of observations	*531*			
ρ^2	*0.6002*			
Log likelihood	*−233.2257*			

Notes: *** significant at 1% level; ** significant at 5% level; * significant at 10% level.
Source: GRPI, Ethiopian survey, 2006/2007

failure. Therefore, they are very risk averse and less inclined to take up non-adaptable and/or non-stable teff varieties.

Results in Table 2.8 also showed that the demand for environmental adaptability attribute of teff varieties increased with the household size. The shock to output associated with growing non-adaptable varieties has a much larger negative effect on larger households than smaller ones, inducing bigger households to be more risk averse towards such crops.

Conclusions and policy implications

The aim of this study was to estimate the private values that farmers attached to crop variety traits and to identify the most important farm household contextual factors that condition their variety attribute preferences. Data was collected in personal interviews from sorghum and teff growing farmers in two peasant associations (PAs) of North Wollo zone. The choice experiment method was applied to investigate farmers' demand for crop varieties and their attributes conditional on the characteristics of the households and the main decision-makers.

The results of both measures of MWTP (with producers' price and productivity taken as the monetary attributes alternatively) revealed farmers' strong preferences for environmental adaptability for both teff and sorghum. Yield stability was also more important than increased productivity. These findings may explain the low adoption rates of high-yield variety seeds in Ethiopia over the last several decades.[9] The fact that farmers attached sizeable values to both environmental adaptability and yield stability traits of sorghum and teff points to the need for supplying a crop genetic variety with additional attributes of resilience to harsh environmental conditions, rather than a breeding strategy that solely targets enhanced agricultural productivity. The results also revealed that there were differences among farm households in terms of household characteristics, resource endowments, extension participation and off-farm job opportunities that affect farmers' private valuations of crop variety traits. There were significant differences between farmers that manage larger and smaller lands, between experienced and less experienced farmers, and between households with low and high values of livestock.

These results have important implications in the areas of on-farm conservation and variety adoption. First of all, farm households who attached the highest values to attributes already embedded in traditional varieties would maintain the varieties *de facto*. Targeting these farmers would minimize conservation costs and enhance compliance in on-farm conservation activities. For instance, *de facto* conservation of environmentally adaptable sorghum varieties by more experienced farmers with small land areas implies that there is little need to design external incentives for these varieties. This strategy, however, needs close follow-up and is likely to change in the medium to long run with farmers' incentives. For instance, the transformation of Ethiopia's rural infrastructure such as roads and markets that is occurring in the country will increasingly provide farmers with the incentive to shift from environmentally adaptable and stable yielding varieties towards highly productive and commercial crops. In such instances external incentives will have to be in place to ensure on-farm conservation of these crops.

Second, understanding farmers' variety trait preferences also informs decision-makers about the variety attributes that have to be considered in on-farm conservation. For instance, more experienced farmers and small farm holders with smaller livestock assets were affected the most when they had to forego teff and/or sorghum varieties with better yield stability and environmental adaptability. They are, therefore, less likely to cooperate with on-farm conservation activities that expect them to replace varieties with these attributes unless they get equivalent compensation.

The third important policy implication relates to the area of variety adoption. For agricultural technologies to be successful, their attributes should address farmer concerns. Clearly, understanding farmers' variety trait preferences is an input to this end. For instance, according to the results, to target and address variety demand for asset-poor, experienced and larger farmers, the priority variety attributes are environmental adaptability and yield stability of both teff and sorghum varieties.

Notes

1   A peasant association (PA), often comprised 400 to 500 people, is the smallest rural unit in the government's organizational structure in Ethiopia.
2   Respondents were asked to specify the walking distance (in minutes) for each type of infrastructure and then an average walking distance (in minutes) was calculated for each household.
3   Birr is Ethiopia's currency where Ethiopian Birr 8.93 = US$1 at the time of the experiment (June and July of 2007).
4   Even though environmental adaptability and yield stability are linked, we separated them because a 'non adaptable' variety can still be conceived to give 'stable yield' in the absence of drought and frost problems. In designing choice experiments, one assumes that the alternatives are mutually exclusive (Hensher et al, 2005) while the attributes need not be mutually exclusive. Some level of inter-attribute correlation is unavoidable, which is the case in this study for yield stability and environmental adaptability attributes.
5   The number of crop varieties that can be generated from 4 attributes, 2 with 3 levels and the remaining 2 with 2 levels is $3^2 * 2^2 = 36$.
6   Fractional factorial designs or main effects involve the selection of a particular subset or sample (i.e. fraction) of complete factorials (possible combinations), so that particular effects of interest can be estimated as efficiently as possible (Louviere et al, 2000).
7   This procedure makes the variations of the attributes of the crop descriptions (profiles) uncorrelated in all choice sets (Alpizar et al, 2001).
8   After running a Pearsonian bivariate correlation between household size and sex of the household head, it was found that the two variables are positively and significantly correlated (0.01 significance level).
9   Despite huge investments and extension programmes to promote improved seeds, the use of improved seeds is still very low – only 3 to 5 per cent of Ethiopia's cultivated agricultural area is covered with improved seeds – leaving a great proportion of the farm households to depend on traditional varieties (World Bank, 2005).

References

Alpizar, F., Carlsson, F. and Martinsson, P. (2001) 'Using choice experiments for non-market valuation', *Economic Issues*, vol 8, no 1, 83–109

Bateman, I. J., Carson, R. T., Day, B., Hanemann, W. M., Hanley, N., Hett, T., Jones-Lee, M., Loomes, G., Mourato, S., Ozdemiroglu, E., Pearce, D. W., Sugden, R. and Swanson, S. (2003) *Guidelines for the Use of Stated Preference Techniques for the Valuation of Preferences for Non-market Goods*, Edward Elgar, Cheltenham

Bezabih, M. (2008) 'Agro-biodiversity conservation under an imperfect seed system: the role of community seed banking schemes', *Agricultural Economics*, vol 38, no 1, pp77–87

Birol, E., Smale, M. and Gyovai, Á. (2006) 'Using a choice experiment to estimate farmers' valuation of agrobiodiversity on Hungarian small farms', *Environmental and Resource Economics*, vol 34, no 4, pp439–469

Breffle, W. S. and Morey, E. R. (2000) 'Investigating preference heterogeneity in a repeated discrete-choice recreation demand model of Atlantic salmon fishing', *Marine Resource Economics*, no 15, pp1–20

Bunch, D., Louviere J. J. and Anderson, D. (1996) 'A comparison of experimental design strategies for choice-based conjoint analysis with generic-attribute multinomial logit models', Working paper, Graduate School of Management, University of California, Davis, CA

Carlsson, F., Fraykblom, P. and Lagerkvist, C. J. (2007) 'Consumer benefits of labels and bans on GM foods-choice experiments with Swedish consumers', *American Journal of Agricultural Economics*, vol 89, no 1, pp152–161

di Falco, S., Bezabih, M. and Yesuf, M. (2010) 'Seeds for livelihoods: crop biodiversity and food production in Ethiopia', *Ecological Economics*, no 69, pp1695–1702

Freeman III, A. M. (2003) *The Measurement of Environmental and Resource Values: Theory and Methods*, 2nd edition, Resources for the Future, Washington, DC

Greene, W. (2000) *Econometric Analysis*, 4th edition, Prentice-Hall, Upper Saddle River, NJ

Hanley, N., Wright, R. E. and Adamowicz, W. L. (1998) 'Using choice experiments to value the environment', *Environmental and Resource Economics*, vol 11, no 3–4, pp413–428

Hausman, J. A. and McFadden, D. (1984) 'Specification tests for the multinomial logit model', *Econometrica*, vol 52, pp1219–1240

Hensher, D. A., Rose J. and Greene W. (2005) *Applied Choice Analysis: A Primer*, Cambridge University Press, Cambridge, UK

Lancaster, K. (1966) 'A new approach to consumer theory', *Journal of Political Economy*, no 74, pp132–157

Louviere, J. J., Hensher, D. A., Swait, J. D. and Adamowicz, W. L. (2000) *Stated Choice Methods: Analysis and Applications*, Cambridge University Press, Cambridge, UK

Luce, D. (1959) *Individual Choice Behaviour*, John Wiley, New York, NY

McFadden, D. L. (1974) 'The measurement of urban travel demand', *Journal of Public Economics*, no 3, pp303–328

Naeem, S., Thompson, L. J., Lawler, S. P., Lawton, J. H. and Woodfin, R. M. (1994) 'Declining biodiversity can affect the functioning of ecosystems', *Nature*, no 368, pp734–737

Omondi, I., Baltenweck, I., Drucker, A. G., Obare, G. and Zander, K. K. (2008a) 'Valuing goat genetic resources: a pro-poor growth strategy in the Kenyan semi-arid Tropics', *Tropical Animal Health and Production*, no 40, pp583–596

Omondi, I., Baltenweck, I., Drucker, A., Obare, G. and Zander, K. K. (2008b) 'Economic valuation of sheep genetic resources: implications for sustainable utilization in the Kenyan semi-arid Tropics', *Tropical Animal Health and Production*, no 40, pp615–626

Rolfe, J., Bennett, J. and Louviere, J. (2000) 'Choice modelling and its potential application to tropical rainforest preservation', *Ecological Economics*, vol 35, no 2, pp289–302

Ruto, E. and Garrod, G. (2009) 'Investigating farmers' preferences for the design of agri-environment schemes: a choice experiment approach', *Journal of Environmental Planning and Management*, vol 52, no 5, pp631–647

Scarpa, R., Drucker, A., Anderson, S., Ferraes-Ehuan, N., Gomez, V., Risopatron, C. N. and Rubio-Leonel, O. (2003a) 'Valuing animal genetic resources in peasant economies: the case of the Box Keken Creole Pig in Yucatan', *Ecological Economics*, vol 45, no 3, pp427–443

Scarpa, R., Kristjanson, P., Ruto, E., Radeny, M., Drucker, A. and Rege, E. (2003b) 'Valuing indigenous farm animal genetic resources in Kenya: a comparison of stated and revealed preference estimates', *Ecological Economics*, vol 45, no 3, pp409–426

Smale, M. (ed.) (2006) *Valuing Crop Biodiversity: On-farm Genetic Resources and Economic Change*, CABI Publishing, Wallingford, UK

Train, K. (2003) *Discrete Choice Methods with Simulation*, Cambridge University Press, Cambridge, UK

Wale, E. (2004) 'The economics of on-farm conservation of crop diversity in Ethiopia: incentives, attribute preferences, and opportunity costs of maintaining local varieties of crops', PhD thesis, University of Bonn, Bonn, Germany

Wale, E. and Yalew, A. (2007) 'Farmers' variety attribute preferences: implications for breeding priority setting and agricultural extension policy in Ethiopia', *African Development Review*, vol 19, no 2, pp379–396

World Bank (2005) *Ethiopia – Well-being and Poverty in Ethiopia: The Role of Agriculture and Agency*, Report no 29468-ET, World Bank, Washington, DC

Zander, K. K. and Drucker, A. G. (2008) 'Conserving what's important: using choice model scenarios to value local cattle breeds in East Africa', *Ecological Economics*, no 68, pp34–45

Zander, K. and Holm-Mueller, K. (2007) 'Valuing farm animal genetic resources by using a choice ranking method', in J. Meyerhoff, N. Lienhoop and P. Elsasser (eds) *Stated preference methods for environmental valuation: applications from Austria and Germany*, Metropolis, Marburg, Germany

Chapter 3

Valuation of Rice Diversity in Nepal: A Trait-based Approach

Krishna Prasad Pant, J. C. Gautam and Edilegnaw Wale

Summary

Commercialization of valuable consumption traits (like aroma, taste and easy expansion in cooking) and production traits (like high yield and pest resistance) can make traditional crop varieties more attractive for local farmers, enhancing their on-farm conservation. Market-driven methods of conservation based on incentives and opportunity costs require *a priori* knowledge about farmers' preferences for varieties and traits. This case study has attempted to value different useful traits of rice landraces grown in Nepal. A sample of randomly selected 200 Nepalese rice growers, 100 each from the hills and the plain, were surveyed on production of rice landraces and the market price fetched by each of them.

Two types of valuation methods were used: hedonic pricing and contingent valuation. The results of the hedonic pricing method (HPM) showed that consumers value aromatic and tasty traits of rice landraces close to NPR (Nepalese rupee) 11 billion ($148.6m) and NPR 2 billion ($27m), respectively. The contingent valuation method (CVM) was employed for estimating farmers' derived demand for hypothetical seeds with different useful traits combined as desired by the farmers. The results showed that farmers were willing to pay nearly NPR 1 billion (close to $13.5m) for high-yielding landraces with aromatic traits and over NPR 1 billion for disease-resistant landraces highly suitable for cooking. These values of unique traits of rice landraces are likely to exceed the costs of conservation. The estimates are indicative of the values of the rice traits embodied in the rice landraces that justify the need for their conservation. Therefore, it has been concluded that every dollar spent in conservation of such landraces makes the society better-off.

Background

Biodiversity on the earth is a reservoir of genetic resources (GRs) that have been used by humans for centuries and have vast potential for the production and manufacture of food, pharmaceutical and cosmetic products for the generations to come. Though biodiversity exists at the genetic (allelic variation), species and ecosystem levels (CBD, 1992), genetic level diversity is more important for food and agriculture as they are important for future crop breeding with conventional technology as well as modern biotechnology. The diversity of food plants consists of crop resources that are created and maintained as active components of agro-ecosystems (Brookfield and Padoch, 1994; Vandermeer et al, 1998) by the farmers.

As noted in Chapter 1, the need for the conservation of the biodiversity is indispensable. Conservation of GRs is important for future production of food that is needed for the sustenance of the human race on the earth. The literature suggests that the continued production of agro-biodiversity is dependent upon adequate supplies of farm resources among rural households (Brush et al, 1992; Mayer and Glave, 1999). A wealth of indigenous knowledge associated with the utilization of plants and animals exhibits the food and medicinal values of the GRs to local communities, including consumers, producers and other actors involved in the market value chain. Many of the farmers in developing countries are joint producers and consumers of the food, and both consumption traits and production traits of the crops are relevant for them.

Nepal has developed many modern varieties that, as expected, give higher yield in shorter duration than the landraces.[1] There is a high rate of replacement of landraces by the modern varieties. The higher the profit gap from the modern varieties and landraces, the faster the replacement process. Thus, finding an effective strategy for conservation and sustainable utilization of crop GRs would involve enhancing the comparative advantage of the landraces, i.e. reduction in the profit from the modern varieties and/or increase in the profit from the landraces. The decrease in the profit from the modern varieties is not desired as it decreases the welfare of the people. That makes increasing the profitability of landraces the most plausible mechanism to improve their maintenance and slow down the replacement. Promoting the commercial use of GRs for increased profit to the farmers who grow them is emerging as one of the major strategies of effective *in situ* agro-biodiversity conservation.[2] This approach of agro-biodiversity conservation is discussed in Chapter 6 in detail.

Sustainable use and conservation through commercial use of landraces builds on farmers' self-interest and it is incentive-based rather than the 'command and control' approach. It is more suited to agro-biodiversity conservation that has larger direct use values on-farm. More specifically, the commercialization of GRs can generate income for the farmers and let them realize the economic importance of the resources for their livelihoods. However, the origin of new conflicts for dealing with biodiversity stems from the rules of division and appropriation of the benefits out of the commercial use of the genetic resources. As policy-makers cannot fully understand the value of the landraces, it is likely that they are unable to negotiate with giant seed development companies (including biotechnology companies) for the best use of GRs and their effective conservation.

The future prospects of commercialization, however, crucially depend on the potential market value of the GRs. The market value, in turn, depends on three factors, namely:

1   How much one can commercialize it.
2   How much the breeding/seed development industry is willing to pay for the samples of GRs.
3   How much revenue a single provider can earn.

For GRs that have many potential suppliers and few seed development companies to buy them, the suppliers cannot expect revenue enough for their conservation as the market has a monopsonistic nature. Whether or not a market for GRs can effectively support the conservation of biodiversity essentially depends on the scarcity of the GRs. It is often assumed that the scarcity of the GRs, particularly those that control commercially important traits (such as cooking and production qualities), is rising as demand for them increases due to current advances and future prospects in the seed development sector.

Some crop GRs are found to be owned, managed and used by a few farmers or a limited number of communities. To the extent that benefits accrue to their users and owners, crop GRs are private goods. Their benefits are also public to the extent that they are accrued by all economic agents involved in the market value chain. Due to its public goods nature, the use of biodiversity by one person does not exclude others from using it. Therefore, as noted in Chapter 1, agro-biodiversity has both public and private goods features.

Replacement of the indigenous varieties by exotic high-yielding varieties and changes in farming practices and land use patterns are important causes of agro-biodiversity loss in Nepal (Upreti and Upreti, 2002). This is a typical case of market failure as the farmers fail to value the benefits to society from the conservation of the landraces. As noted in Chapter 1, farmers always maintain crop GRs to the extent that these resources address their household concerns. Thus, their conservation is not optimal as there are crop varieties that have little utility to address farmers' current concerns but have potential future public utility. Market failures, particularly its failure to account for the public goods values, are the major causes of loss in agro-biodiversity. Estimating the total economic value (TEV)[3] of landraces by means of non-market valuation methods can help to develop policies that address the problem of market failures. The valuation of biodiversity can assist decision-makers with the development of mechanisms of equitable sharing of benefits from utilization of GRs, and help to justify investments in their conservation.

As noted in Chapter 1, environmental values can be estimated using revealed and stated preference methods. The revealed preference methods use the market data as revealed by the respondent (consumer, farmer, etc.) while the stated preference methods are based on the preferences stated by respondents under hypothetical market situations. The discussion of stated preference methods in resource and environmental economics dates back to the 1940s (Ciriacy-Wantrup, 1947). Extension of these valuation techniques to agricultural biodiversity is, however, a recent phenomenon (see e.g. Hoyos, 2010).

This chapter aims to evaluate the genetic diversity of traditional rice varieties in Nepal, using both revealed and stated preference methods. The outcome of this case study is expected to help in designing cost- and benefit-sharing approaches for the conservation and use of rice GRs. Information about the values of traditional rice varieties will be valuable to prioritize their on-farm conservation. Hedonic pricing method was used to find the value given by consumers for each trait of rice landraces and contingent valuation method was used to estimate the bidding price of a new variety of rice with useful desirable traits required by farmers.

Approaches for measuring economic value of genetic diversity

Human decision-making (in natural resource use and agricultural technology adoption) involves a series of trade-offs such as between environmental concerns (e.g. biodiversity protection) and meeting the immediate economic needs (e.g. income generation and food security). If traditional crop varieties are low yielding, on-farm conservation of these resources will involve opportunity costs in terms of food production and productivity. Economic principles of valuation can offer mechanisms to estimate the value of agro-biodiversity and make sound decisions aimed at internalizing the trade-offs, while contributing to both objectives of agro-biodiversity conservation and enhancing agricultural productivity.

Measuring the value of biodiversity is a great challenge. Reid et al (1993) observed that even though the debates on the measurement of biodiversity started in the 1950s, there is still no clear consensus about how the value of biodiversity should be measured. Pearce and Moran (1994) examined some aspects of measurement of biodiversity at genetic, species and ecosystem levels. Accordingly, the genetic differences can be measured in terms of phenotypic traits, allelic frequencies or deoxyribonucleic acid (DNA) sequences. The measurements of allelic diversity and DNA sequence require high-level technical information which is out of the scope of this paper. This study relies on the phenotypic characteristics of the products. The consumers and farmers (i.e. the users) can observe the consumption and production characteristics of rice at the grocery store and decide which ones to buy. Most of the characteristics can only be known after the product has been used as a consumer good and/or production input. For example, the consumption characteristics of rice will be known after cooking and eating whereas the production characteristics of the seed will be known to the farmers during their experience with the seed from its storability as an input all the way through post-harvest traits (germination, early maturity, pest/disease resistance, tolerance to water stress, yield, marketability, perishability, post-harvest loss, etc.).

In this case study, rice consumers are surveyed to capture their preferences for different combinations of consumption traits of rice and the demand for seeds by the farmers is taken as the derived demand. Use values are assumed to be the most important part of the values of rice GRs from a consumer point of view and, by applying hedonic pricing, these use values can be assessed. The application of the contingent valuation method, on the other hand, allows the assessment of the TEV of rice GRs.

Methodology

The sources of data, sampling designs and analytical procedures underlying the empirical analysis are discussed in this section.

Data and sources

For valuation of GRs, two districts, namely, Kaski from the hills and Bara from Terai (the plains at the base of the Himalayas), were selected purposively in consideration of the richness of rice diversity. From each district, two village development committees (called villages hereafter) with a high concentration of the rice landraces have been identified by a survey of key informants including professionals, researchers in agro-biodiversity and extension officers in agriculture, all working in the district.[4] Key informants were asked to rank five villages in the district with the highest diversity of rice landraces. Data were collected on accessibility (to markets, roads and transport services), area characteristics and rural development interventions in place. The area characteristics include the number of farm households, total geographic area, distance from district head quarter, area under paddy and total irrigated area. These statistics were scored and aggregated. The two villages with the highest aggregate scores were selected for the survey. On this basis, Lekhnath and Lumle villages were selected from Kaski district (Begnas area) and Kacharba and Maheshpur villages were selected from Bara district (Kacharwa area).

Considering the situation of the villages and the farm households, a checklist of questions for focus group discussions was prepared. From the discussions with the relevant stakeholders, a draft questionnaire for household surveys was drawn. More focus group discussions were conducted to refine the draft questionnaire. The questionnaire was then pre-tested on 30 farm households in each survey district.

A household was the sampling unit. The farm households cultivating the rice landraces formed the sampling frame. The sample households were selected using a simple random sampling method without replacement. A sample of 50 farm households was drawn from each of the four selected villages. Thus, 200 sampled households were surveyed altogether for estimating the value of

Table 3.1 *Relative abundance of rice landraces and varieties in the farmers' fields*

Relative abundance (%)+	Name of the Variety	No of varieties
41 to 50	BG-1442*	1
31 to 40	Basmati	1
21 to 30	Sona Masuli* and Anadhi	2
10 to 20	Kathe, Anga, Meghdoot, Jetho Budho, Pahele, Mansuli*, China-4*, Sabitri*, Sotwa, Rekshali, Ekle, Dhudhraj and Harinkar	13
Less than 10	Rate, Lumle-2*, Kalopatle, Chaite-1*, Gurdi, Biramful, Chhatraj, Mutmur, Rato Anadi, Bayarni Jhinuwa, BGAR-4*, Chhote, Ghaiya-2*, Jerneli, Mansara, Seto Anadhi, Darmali, Madhesi, Manamuri, Natwar, Radha-7*, Budho Sigdeli, Janaki*, Nakhisaro, Radha-9*, Barkhe-2*, Kaskeli Thude, Gauria, Kathe Gurdi, Ranga, Sokan, Thulo Gurdi, Machhapuchhre, Khumal-4*, Chhomrong, Sathi, Gajale, Gurdi, Jaya, Madhumala, Masula, Philipes, Lamjunge, Deurali, BiramfulxHimali*, Ekle hybrid, EklexKY*, HY-6264*, IR-6465*, Mansara Hybrid*, Sano Gurdi, Sano Gurdi Hybrid*, Sano GurdixNR*, Seto Gurdi, Thulo Gurdi Hybrid*, Thulo gurdixNR*, Rato Darmali, Deupure Kathe, Kaskeli Kathe, Bhelasaro, Lahare Gurdi and Bhalu	61
	Total	78

Notes: +The relative abundance is measured as the percentage of the households growing that variety. The list of the varieties and landraces are in the descending order of relative abundance, BG-1442 being the most abundant and Bhalu being the least. The asterisks '*' at the end of the name of the varieties designate modern rice varieties and those without indicate a landrace.
Source: 2006 household survey, Nepal

important traits of rice varieties. A team of two well-trained research assistants was deployed for the survey. The pre-tested questionnaire was administered by trained enumerators.

According to the survey data, the price of rice varieties varies from NPR 725 ($10)[5] (for Chaite and Janaki varieties) to NPR 1622 ($22) (for Basmati varieties). The sampled 200 farmers were found growing 78 rice varieties altogether. Nearly 50 per cent of the farmers were found growing a high-yielding variety (BG-1442). Over 30 per cent of them were growing Basmati rice. Four varieties, namely BG-1442, Basmati, Sona Masuli and Anadhi, were the most popularly grown varieties. Most of the landraces were grown by less than 10 per cent of the farmers. If one is to take relative abundance and continuous use of traditional rice varieties as an index for on-farm conservation, those landraces grown by a small number of households are more likely to be lost easily if some conservation measures are not applied. The relative abundance of each of the variety is presented in Table 3.1.

Farmers' knowledge of different traits of rice varieties can be used as an input in the sustainable use and conservation of agro-biodiversity. Their knowledge is the accumulation of the inherited experiences, updated by the observations in the field (Berkes et al, 1995; de Boef, 2000). They understand the connection between organisms and their surrounding environment (Perrings et al, 1995; Wood and Lenne, 1999). As a result, they grow different landraces of rice that are better suited to their farming systems (taking the seeds as production input) and their cooking preferences (taking rice as consumption good).

Hedonic pricing model

One of the widely used environmental valuation methods as a subset of the revealed preference approach is the hedonic pricing method. The philosophy behind hedonic pricing is that people pay for a product by valuing the embedded bundles of attributes of this product. The conceptual and analytical basis of this valuation technique emanate from Lancaster's characteristics model (Lancaster, 1966). The early applications of this method, however, started in the 1920s. The first application of hedonic modelling is found in fresh vegetables. Though Waugh (1928) first used hedonic pricing on land characteristics, Ridker (1967) was the first to use this method on environmental goods for estimating the marginal value of air quality in residential areas.

Rosen (1974) used the hedonic price theory to interpret the derivative of hedonic property price function with respect to air pollution as a marginal implicit price and, therefore, the marginal willingness-to-pay (MWTP) of individuals for air pollution reduction. Rosen's model starts with a distribution of utility-maximizing buyers and a distribution of profit-maximizing sellers. The equilibrium is achieved when the variation in price reflects the variation in the attributes under the condition of full information. Price (p) of a house, say, with a vector of attributes (z) and a vector (α) of parameters (Haab and McConnell, 2002) can be written as:

$$p = h(z, \alpha)$$

(Eqn 3.1)

The equilibrium will exist when the buyer maximizes the utility from consumption of a composite bundle of commodities (x) with a vector of household preference function (β). The utility function is $u(x, z, \beta)$ and budget constraint with income $y = h(z) + x$. Maximizing the utility subject to the budget constraint, we get optimal condition for each attribute.

$$\delta u(x, z, \beta)/\delta z_i = \lambda \, \delta h(z)/\delta z_i, i = 1, ..., n$$

(Eqn 3.2)

where the Lagrangian multiplier, λ, is the marginal utility of the income. From Equation 3.2, MWTP for the i-th attribute can be calculated as $\delta h(z)/\delta z_i$.

The hedonic price method has emerged as a powerful tool for the valuation of environmental goods that can be extended for valuation of newly recognized environmental amenities like biological diversity.

Dalton (2003) estimated a nonseparable household hedonic pricing model of upland rice attributes combining both production and consumption traits. Accordingly, yield was not a significant attribute in determining farmers' WTP for new varieties. However, this trait has served as the defining factor for promoting a new variety for official release. What amount the farmers are paying for a particular new variety that comprises different gene combinations can be taken as the revealed preference of the farmers for genetic resources.

The value of major rice traits established by the Nepalese market can be estimated using hedonic pricing method based on the market price of varieties. It is assumed that the price the farmers pay for the rice depends on the variety attributes of rice, i.e. consumers pay more for the more useful variety attributes. Using this principle of price determination on the basis of the attributes of the product, the following model is used for estimating the use value established by the consumers on rice that are attributable to major traits of rice landraces.

$$PP = \alpha + \beta_1 T + \beta_2 A + \beta_3 BR + \beta_4 LS + \beta_5 Md + \beta_6 Ce + \beta_7 Ex + \beta_8 ST + \beta_9 MP + \beta_{10} TA + \beta_{11} Ms + v$$

$$\text{(Eqn 3.3)}$$

where PP, T, A, BR, LS, Md, Ce, Ex, ST, MP, TA, Ms and v refer to the price of paddy rice, tasty trait, aromatic trait, quality for bitten rice,[6] quality for 'latte' and 'siroula',[7] medicinal uses, uses in ceremony, expansion in cooking, good storage quality, milling per cent, Terai area, season and the error term respectively (a more complete description of all these variables is given in Table 3.3). β_1 to β_{11} are coefficients to be estimated.

Contingent valuation method

Revealed preference methods of environmental valuation are preferred over the stated preference methods like the contingent valuation method (CVM). However, the revealed preference methods cannot always be employed, particularly when the values of the resources have not yet been realized. The diversity of landraces has immense future potentials from breeding high-yielding varieties with useful traits available in the landraces. Therefore, the potential future values of those varieties can be valued using only stated preference methods. The CVM relies on a questionnaire survey about WTP of individuals for conservation of a certain environmental resource. Pearce and Moran (1994) have argued that CVM is a promising option for biodiversity valuation in general because of the potential for information provision and exchange during the survey process, which offers scope to experiment with respondent knowledge and understanding of biodiversity. A variation of this approach has been used by Brown and Goldstein (1984) in order to value *ex situ* (plant) collections. They used a model where the benefits of reducing expected future production losses are weighed against gene-bank operating costs and searches, arguing that all varieties should be conserved for which the marginal benefit of preservation exceeds marginal cost. Oldfield (1989) focuses on actual crop losses (in this case related to southern corn leaf blight) as a measure of value of the genetic improvement efforts used to eventually overcome such losses. A recent study by Poudel and Johnsen (2009) has also applied CVM to assess rice landraces in Nepal.

In this study, two basic steps were followed to estimate the value of useful traits in the landraces:

1   A hypothetical description (scenario) of the new rice varieties, with combinations of different traits available in landraces, was presented to the farmers. This included the combination of different useful traits into a single variety. The farmers were asked to bundle the different useful traits of rice into a single variety.

2   The farmers were asked questions to determine how much they would be willing to pay for 1kg of seed of the new variety combining the useful traits they have chosen. These questions took the form of asking how much a farmer was willing to pay for some new variety that contains a desired mix of the useful traits. Depending on the preferred elicitation format, econometric models are then used to infer a WTP for the change. An aggregate welfare measure was calculated by multiplying the mean with the relevant population of users.

The values put by the farmers on the rice seeds with different traits were estimated based on farmers' bid for a price of a hypothetical rice variety with a combination of traits they desire to have in a single variety. The following mathematical model was used for empirical estimation of the value the farmers attach to different combinations of the traits.

$$PS = \alpha + \beta_1 HA + \beta_2 T + \beta_3 Ex + \beta_4 SL + \beta_5 DrR + \beta_6 LF + \beta_7 DsR + \beta_8 TA + \beta_9 A + \beta_{10} FR + \beta_{11} F + \beta_{12} PT + \beta_{13} RA + \beta_{14} I + v$$

(Eqn 3.4)

where *PS, HA, T, Ex, SL, DrR, LF, DsR, TA, A, FR, F, PT, RA, I* and *v* refer to price of hypothetical seed, high-yielding and aromatic traits, tasty trait, cooking expansion trait, suitability to be sown late, drought resistance, suitability for less fertile land, disease resistance, Terai area, age of the household head, gender, number of family members, number of plots of paddy land, ownership of radio, non-agriculture income and the error term respectively (a more complete description of all these variables is given in Table 3.5). β_1 to β_{14} are coefficients to be estimated.

Table 3.2 *The most popular four varieties*

Variety	Type	Relative abundance (%)	Average area (ha)	Productivity (qt*/ha)	Price (NPR/qt*)	Gross return (NPR/ha)
BG-1442	Improved	44.5	0.26	41	1074	44,034
Basmati	Landrace	37.5	0.16	30	1622	48,660
Sona Masuli	Improved	29.5	0.23	49	1126	55,174
Anadhi	Landrace	26.5	0.02	14	1168	16,352

Note: * qt stands for quintal, 1qt = 100kg.
Source: 2006 household survey, Nepal

Estimation of the value consumers put on rice landraces

Farmers planted each variety separately on different plots or sometimes in different parts of the same plot. On average, each farmer grew 4.62 varieties of rice every year, each on small areas. For example, the average area under the most popular variety (grown by over 44 per cent of the households), modern variety BG-1442, was 0.26ha followed by the landrace Basmati 0.16ha (Table 3.2). Though Basmati was grown by a larger proportion of the households (nearly 38 per cent) than Sona Masuli (nearly 30 per cent), the area commanded by these two varieties showed the opposite. This was more distinct in the case of Anadhi. Though this long grain aromatic landrace was grown by over one-fourth of the total households, the average area under this landrace per household was very small. These precious landraces (like Basmati and Anadhi) are unable to compete with other commercially grown modern varieties. They are thriving in the farmers' field only due to their special and desirable phenotypic characteristics.

The reason for the small area of land under Anadhi was its low price and productivity. It is clear that the gross return from Anadhi is about one-third of the gross return from the competing modern rice varieties. It showed that the typical landraces with unique genes are unlikely to survive *in situ* under business as usual situations. The options are either to go for *ex situ* conservation or to make the society understand the value of these landraces and find market-based *in situ* conservation mechanisms to benefit farmers and generate better consumption outcomes.

The descriptive statistics of rice traits used for the hedonic pricing model are presented in Table 3.3. Though some of the traits appeared to be not mutually exclusive, they were used separately as long as there was no problem of multi-collinearity.

These variables were fitted to a regression model to estimate the contribution of different traits to the price of the type of rice purchased. Many phenotypic properties like taste, aroma, expansion in cooking, storage quality and suitability to various dishes (bitten rice, *latte* and *siroula*, and ceremonial dishes), and

Table 3.3 *Descriptive statistics of rice traits and price (n = 932)*

	Variable name	Variable description	Mean	Standard deviation	Expected sign
1	Tasty	Dummy variable showing presence of tasty trait (1 if tasty and 0 otherwise)	0.61	0.49	+
2	Aromatic	Dummy variable showing presence of aromatic trait (1 aromatic and 0 otherwise)	0.19	0.39	+
3	Bitten	Dummy variable for bitten rice (1 good for bitten rice and 0 otherwise)	0.19	0.39	+
4	LateSiro	Dummy variable for *Latte* and *Siroula* (1 good for Latte and Siroula and 0 otherwise)	0.08	0.27	+
5	Medicine	Dummy variable for medicinal properties (1 good for medicinal uses and 0 otherwise)	0.10	0.30	+
6	Ceremony	Dummy variable for special ceremonies (1 good for special ceremonies and 0 otherwise)	0.18	0.39	+
7	Expansion	Dummy variable for expansion (1 good for expansion in cooking and 0 otherwise)	0.20	0.40	+
8	Storage	Dummy variable for storability (1 good for storability and 0 otherwise)	0.28	0.45	+
9	Milling	Percentage of rice flour recovered in milling	62.43	5.82	+
10	Terai	Dummy variable for geographic area (1 for plain area and 0 for hill)	0.48	0.50	−
11	Season	Dummy variable for season (1 for main season and 0 for summer season)	0.92	0.27	+
12	PadyPric	Price of paddy rice (NPR per 100kg)	1033.87	278.54	

Source: 2006 household survey, Nepal

medicinal values were preferred by the consumers. These results were expected *a priori.*

Some undesirable traits were also identified. The undesirable traits included coarse grain (beside medium grain) and lack of taste (beside medium taste). The undesirable traits were hypothesized to bear negative prices in the bundle of properties. As the price was taken for fresh harvest of paddy rice, the milling percentage was also a concern for the buyer. It was assumed that the higher the milling percentage, the higher the amount the buyer would pay, keeping all other traits constant (all else being equal). The geographical area 'Terai' was fitted to catch the fixed effects of hills and plains.

Table 3.4 *Factors affecting the price of paddy rice in Nepal (the results of HPM)*

	Variable	Coefficients	Standard error	95% confidence interval	
I	Tasty	36.55**	17.75	1.72	71.37
2	Aromatic	293.44***	21.20	251.84	335.03
3	Bitten	−50.65***	18.72	−87.38	−13.92
4	LateSiro	122.20***	29.00	65.29	179.11
5	Medicine	48.24*	28.94	−8.55	105.04
6	Ceremony	124.80***	21.52	82.57	167.03
7	Expansion	−2.84	19.63	−41.37	35.69
8	Storage	15.13	17.62	−19.45	49.71
9	Milling	−5.68***	1.36	−8.36	−3.01
10	Terai	152.10***	18.76	115.27	188.92
11	Season	85.31***	26.81	32.70	137.92
12	Constant	1129.07***	92.01	948.51	1309.64
	N = 93			$F_{(11, 920)} = 57.99$***	
	Prob > F = 0.000			Adjusted R^2 = 0.402	

Notes: * qt stands for quintal, 1qt = 100kg. *** significant at 1% level; ** significant at 5% level; * significant at 10% level.

Source: 2006 household survey, Nepal

The average price paid by the market to the fresh harvest of paddy was NPR 1034 ($14), ranging from NPR 500 ($7) to NPR 2,400 ($32). If we assume that all farmers obtain equal market opportunities, the variation in the price is due to the difference in the quality, taking quality as the bundle of desirable traits.

The results of the linear hedonic model fitted to the above data are presented in Table 3.4 along with the 95 per cent confidence interval of the coefficients estimated. The model explained over 40 per cent variations in the price of the paddy. To the extent that the model faces omission of the relevant explanatory variables, the marginal value of each trait was overstated. It is also important to bear in mind that the non-use values of rice landraces are not included in this analysis.

The estimates showed that consumers paid NPR 36 ($0.5) per quintal for a tasty trait. This coefficient was on average about 3.5 per cent of average price of paddy. The estimate was highly significant. It can be inferred that by conserving the landrace with this trait and keeping alive the potential of incorporating this trait to other rice varieties in the world, it is possible to maintain the potential benefits of increasing the value of global rice production by over 3 per cent.

Similarly, for the aromatic trait, the consumers were willing to pay NPR 293 ($4) per quintal. This was over 28 per cent of the average price of the rice. If we were to lose this trait, the potential financial loss to society would be 28 per cent of the total value of rice produced globally. This is a value large enough to warrant investing in the required conservation measures.

Rice varieties suitable for bitten rice attracted lower prices (negative coefficient). This finding suggests that this trait had negative impact on utility for consumers relative to other traits. This is because rice varieties just harvested with poor quality and higher moisture content are more suitable for bitten rice. Such varieties are less storable and less preferred for steam rice and hence fetch a lower price. However, the rice varieties with traits suitable for other snacks like latte and siroula can fetch a higher price by NPR 122 per quintal. Similarly, the traits suitable for traditional healing purposes and for use in ceremonies are valued higher than other varieties.

As expected, the varieties with traits that make the rice coarse were valued negatively by the consumers as compared to the medium coarse. The result suggests that the higher the milling percentage, the lower the consumers were willing to pay. This is against *a priori* expectation. This might be because the consumers generally buy milled rice (not paddy rice) and hence the milling per cent is the concern of millers, not consumers. Even if some consumers buy paddy rice to mill it themselves, the milling per cent is not known to the buyers at the time of bidding a price. It is a trait that buyers know through experience. The farmers in the plain region are getting higher price of paddy for similar quality as compared to the farmers in the hill region. This can be because of the fixed effects factors, such as better transportation and communication facilities for better market connectivity in the plain areas.

Rice in Nepal accounts for more than half of the principal food crops. On average, 4 million tons of rice are produced every year (GON, 2007) with an estimated value of NPR 41.4 billion ($559.5m). This means that the aromatic traits of rice generate an extra NPR 11 billion ($148.6m) per annum for Nepal and tasty traits over NPR 2 billion ($27m). However, there are different landraces with different degrees of aroma and different levels of taste. A separate study is worthwhile to quantify and understand the importance of such traits and to value each specific aroma and taste trait.

The analysis apportions the price paid by the consumers to the value given to the different traits. Accordingly, protecting each of the preferred traits roughly increases the value to the society by the respective values.

Estimation of producers' (farmers') values of rice traits

Since it is unlikely that one rice variety will supply all of the attributes that the farmers value, they demand varietal diversity (Joshi and Bauer, 2005). In general, the producers' demand for certain traits is the consumers' derived demand for the traits. Moreover, some traits like high yield and disease (pest) resistance are additional concerns to the producers. If one is to follow the contingent valuation to estimate the producers' valuation of the preferred rice traits, a contingent market has to be created in the minds of the farmers and they would be asked how much they are willing to pay.

Some preferred major traits like aroma, taste, high-milling percentage, expansion in cooking and good storage quality are highly demanded by consumers. In

Table 3.5 *Descriptive statistics of the traits preferred and combined by the farmers*
(n = 600)

	Variable name	Variable description constructed	Mean	Standard deviation	Expected sign
1	YieldAroma	Dummy for high yielding and aromatic traits (1 good for yield and aroma and 0 otherwise)	0.33	0.47	+
2	Tasty	Dummy for taste (1 good for taste and 0 otherwise)	0.23	0.42	+
3	Expansion	Dummy for expansion (1 good for expansion in cooking and 0 otherwise)	0.11	0.32	+
4	SowLate	Dummy for late sowing (1 good even if sown late and 0 otherwise)	0.21	0.41	+
5	Drought	Dummy for drought resistance (1 good for drought resistance and 0 otherwise)	0.18	0.39	+
6	Land	Dummy for soil quality (1 productive under poor soil quality and 0 otherwise)	0.11	0.31	+
7	Disease	Dummy for disease resistance (1 good for disease resistance and 0 otherwise)	0.17	0.38	+
8	Terai	Dummy for geographic areas (1 plain and 0 hill)	0.50	0.50	−
9	Age	Age of the household head (in years)	49.38	9.00	−
10	Gender	Dummy for gender (1 female and 0 male)	0.17	0.38	+
11	Family	Number of family members	7.03	1.98	−
12	PlotPady	Number of plots of paddy land	4.49	2.74	+
13	Radio	Dummy for ownership of a radio set (1 if household owns radio set and 0 otherwise)	0.96	0.20	+
14	Non-farm	Non-agricultural income (in NPR 1000)	79.74	226.94	−
15	Price	Price of the hypothetical seed quoted by the farmers (NPR per kg)	59.69	28.08	

Source: 2006 household survey, Nepal

addition, producers also put value on high-yielding traits, the variety that can be late sown, drought resistance, suitability for less fertile land and disease resistance.

Considering that access to suitable seed is a priority issue for *in situ* agro-biodiversity conservation (Cleveland et al, 1994; Cleveland and Soleri, 2002; Rhoades and Nazarea, 1999; Zimmerer, 2002), the willingness to pay for new type of seeds is analysed. Each sample household was asked to select three combi-

Table 3.6 *Farmers' willingness to pay for seeds of new varieties with different traits (the results of CVM)*

	Preferred traits	Coefficient	Standard error	95% confidence interval	
1	YieldAroma	19.37***	3.35	12.79	25.96
2	Tasty	5.76*	3.32	−0.76	12.27
3	Expansion	22.18***	3.73	14.85	29.51
4	SowLate	−0.83	3.22	−7.15	5.50
5	Drought	0.997	3.33	−5.54	7.53
6	Land	10.74***	3.28	4.30	17.18
7	Disease	21.63***	3.65	14.46	28.799
8	Terai	−21.28***	2.07	−25.35	−17.21
9	Age	0.51***	0.11	0.29	0.72
10	Gender	7.46***	2.67	2.21	12.71
11	Family	−0.32	0.49	−1.28	0.64
12	PlotPady	−0.79**	0.36	−1.49	−0.09
13	Radio	−16.27***	4.99	−26.07	−6.46
14	Non-farm	−0.013***	0.004	−0.02	−0.004
15	Constant	51.43***	8.69	34.36	68.50
N = 600			$F_{(14, 585)} = 22.87^{***}$		
Prob > F = 0.000			Adjusted $R^2 = 0.34$		

Notes: *** significant at 1% level; ** significant at 5% level; * significant at 10% level. The equation explains altogether 34% of the variation in the willingness to pay for the preferred (new) variety of rice.
Source: 2006 household survey, Nepal

nations (X, Y, Z) of desirable traits they prefer in new rice varieties. The traits are considered as fully separable and combinable. The only exception is that the high-yielding and aromatic traits are inseparable for the farmers because the farmers who prefer high-yielding varieties also prefer aromatic ones. The correlation between these two traits is 0.95. The most popular traits selected by the farmers to be incorporated into a new variety are found to be high yield and aroma combined, high milling percentage, taste and a variety that can be sown late (Table 3.5).

To estimate farmers' WTP for different traits, each respondent was asked to bid a maximum price for 1kg of rice seeds with different combinations (X, Y, Z) of desirable traits. Under the condition that the combinations they made are available in the market, the farmers on average are willing to pay NPR 60 ($0.80) per kg of such rice seed.

Farmers' WTP for the hypothetical traits they combined was fitted with the desirable traits they selected, a geographical dummy for catching fixed effects, the age and gender of the respondents, family size, the number of the plots the farmer is operating for rice cultivation, radio set ownership, non-farm income[8] and farmers' quoted price of seed.

The coefficients of different variables that explain farmers' WTP for the rice seed are presented in Table 3.6. The farmers valued NPR 19 per kg extra for high-yielding aromatic traits. The farmers were willing to pay NPR 6 more for tasty trait. One of the major cooking characteristics, the expansion in cooking, was also highly valued by the farmers. For the rice variety that leads to expansion in cooking, the farmers were ready to pay NPR 22 extra for 1kg of improved seed.

The farmers as producers also positively valued the traits that make the landrace suitable for less fertile land. For these seeds, the farmers were willing to pay nearly NPR 11 per kg extra. This reduces the fertilizer cost for the farmers and the environmental problems associated with fertilizer application. For disease-resistant traits, the farmers were willing to pay the most (NPR 22 extra per kg of seed). This reduces the loss of crop and costs of pesticides for the farmers on top of the ecological benefits to the society. In addition, it avoids human health problems due to the pesticide residue they consume with rice and health problems of animals that consume the straw. The result showed the importance of rice genetic resources that can be incorporated to other varieties for the development of new varieties with high-yielding and disease-resistant characteristics.

There were some biases on the part of the respondents, which have been captured. Older farm household heads bid slightly higher prices for the new seed than the younger farmers. Older farmers who knew the traditional food habits and culture could better understand the importance of the traits of landraces than younger farmers. Similarly, female farmers were ready to pay more than male farmers. This might be because the traditional dishes of rice are prepared mostly by women and they are in a better position to understand the values of the rice traits than their male counterparts. The women also have better understanding of their farm situations than men and can appreciate better the production traits of rice.

The farmers in the Terai (plain) areas were willing to pay less for the new varieties. This might be because they had better access to new varieties of rice in border towns of India. Similarly, the larger the number of plots the farmers had, the lower their WTP for the new varieties. This might be because the larger the number of plots they have, the more diverse their agro-ecology and thereby the better their chance to grow a larger number of varieties in their farm. Rana et al (2007) have found that on-farm landrace diversity is positively affected by the number of parcels of land that farmers manage. Unlike previous expectations, the farmers with radio sets (means higher access to the information and a higher living standard) were willing to pay less. This might need further study to disprove or explain it further. A similar result was found for farmers with higher non-farm income. For these farmers, the improved varieties were less important.

The farmers highly valued the production traits and cooking characteristics of rice varieties. Older people in hill areas with a smaller number of plots were willing to pay more for seeds of new varieties with multiple traits than farmers with the opposite features. The farmers, on average, were willing to pay 23 to 62 per cent higher prices for different traits of rice.

In Nepal, more than 1.5 million hectares of land are planted with rice every year (MOAC, 2006). Given that the recommended seed rate for rice is 30kg per ha, the total rice seed planted every year is about 46,500 tons. The farmers were ready to pay NPR 19 per kg extra for seeds of high yielding aromatic traits combined. It means, in total, farmers value NPR 0.88 billion ($11.9m) for high-yielding aromatic traits. Similarly, the farmers on aggregate give a value of NPR 1.03 billion ($13.9m) for each trait of expansion in cooking and disease resistance.

The Green Revolution in Nepal is still considered as a cause for concern through the displacement of landraces by high-yielding and fertilizer-responsive modern varieties. As the market supplies high-yielding modern varieties of rice, the profit-maximizing and risk-taking farmers generally replace the more diverse landraces by more uniform modern varieties. Consequently, a few genetically uniform high-yielding varieties have replaced genetically variable crop landraces over the longer term.

About 53 per cent of the farm households in Nepal continue to grow both modern varieties and landraces side by side. Their demand for both types is clearly shaped by markets, land and soil heterogeneity, and the consumption preferences of their families (Joshi and Bauer, 2006). Other authors have empha-sized the development approaches that can value, conserve, develop and market agro-biodiversity to alleviate the extreme poverty (Bardsley and Thomas, 2005).

To conclude, commercialization of the valuable consumption traits (like aroma, taste and expansion in cooking) and production traits (like yield and pest resistance) can increase the importance of the landraces among the farmers. This approach can help conservation of rice diversity in the farmers' fields.

Conclusions and implications for policy

Understanding the values that users (producers and consumers) put on the specific traits of rice landraces grown by smallholder farmers will be helpful to the design of market-based conservation strategies. With this motivation, this case study has focused on the valuation of genetic diversity of rice landraces in Nepal. The valuation is required not only for developing mechanisms for the equitable sharing of benefits from its utilization but also for the justification of added invest-ment for conservation and bio-prospecting. The study uses both revealed and stated preference methods. Hedonic pricing method was used to find the value given by the consumers for each trait and contingent valuation method was used to value new variety seeds with a combination of traits that they consider useful.

The hedonic analysis apportioned the price paid by the consumers to the value they give to different traits. The study concludes that protecting each of the preferred traits increases their value to society. For instance, the value of the aromatic trait of Basmati or other local landraces can be about one-fourth of the value of rice produced globally.

For Nepal alone, the aromatic traits of rice have values of about NPR 11 billion ($148.6m) and tasty traits over NPR 2 billion ($27m) per annum. These estimates include only the use values of rice that arise from the actual use consist-ing of the direct use value from consumption by the households and option values generated by an individual's WTP to protect rice production against any future risks. The value given by the farmers to use seeds of aromatic landraces was derived from the value given by the consumers. Other use values (like the ecosystem functions of paddy field) and non-use values are also important elements but are not captured here. For all these reasons, the estimates were indications of the lower bounds. As the estimates showed, the values of different unique traits of rice landraces were assumed to be larger than the costs of conservation. The conservation costs will become even more justified and appealing when the total economic values are accounted by including their non-excludable and non-rival use (like ecosystem functions) and non-use values.

As expected, consumers valued more consumption traits (such as aroma and taste) that can maximize their utility. In contrast, farmers as producers valued more production traits (such as high yield and disease resistance) that increase their income.

Market-driven methods of conservation are effective and efficient as they are based on incentives and opportunity costs. Efforts are needed to establish new markets for the conservation of landraces with unique traits. Commercialization of the valuable consumption and production traits can make landraces more attractive for local farmers. This will decrease the income gap between the modern rice varieties and landraces, decreasing the rate of replacement of landraces by the modern varieties. Market-driven methods of conservation are plausible to all actors involved as they are implemented based on incentives and opportunity costs. For both moral and equity reasons, it is essential either to compensate local poor farmers for maintaining low-productive rice landraces or to enhance the comparative advantages of these landraces so that farmers can earn better incomes. From a moral perspective, the poor cannot afford to bear the opportunity costs of agro-biodiversity conservation on behalf of society.[9]

Notes

1   The landraces include the farmers' traditional varieties produced and maintained by them for many generations, and even high-yielding varieties that have been bred and released for more than 15 years have since become incorporated into farmers' own seed production systems (Almekinders and Louwaars, 1999; Cleveland and Soleri, 2002).

2   There are other strategies to protect biodiversity, including *in situ* conservation through protected area conservation, and *ex situ* conservation through zoos, aquaria, botanical gardens, seed banks and gene banks.

3   The concept of TEV states that an environmental good or resource consists of two broad categories of values: use and non-use values (Bateman et al, 2002). Use values are further classified into direct and indirect use values while non-use values consist of existence and bequest values.

4   The key informants include a senior district agricultural development officer, a senior agronomist in the district, a senior extension officer, researchers working on local landraces at the National Agriculture Research Council (NARC), farmers knowledgeable about local agro-biodiversity, and community leaders.

5   Exchange rate May 2010: 1 US$ = 74 NPR.

6   Bitten rice is flattened, dry rice used for snacks.

7   Latte and siroula are local snacks popular on particular occasions.
8   Non-farm income was considered since it affects rice income.
9   The authors duly acknowledge comments and suggestions from anonymous reviewers.

References

Almekinders, C. J. M. and Louwaars, N. (1999) '*Farmers' Seed Production: New Approaches and Practices*', Intermediate Technology, London

Bardsley, D. and Thomas, I. (2005) 'In situ agrobiodiversity conservation for regional development in Nepal', *Geojournal*, no 62, pp27–39

Bateman, I. J., Carson, R. T., Day, B., Hanemann, M., Hanley, N., Hett, T., Jones-Lee, M., Loomes, G., Mourato, S., Ozdemiroglu, E., Pearce, D. W., Sugden, R. and Swanson, J. (2002) *Economic Valuation with Stated Preference Techniques: A Manual*, Edward Elgar, Cheltenham, UK and Northampton, MA

Berkes, F., Folke, C. and Gadgil, M. (1995) 'Traditional ecological knowledge, biodiversity, resilience and sustainability', in C. S. Perrings, K. G. Maler, C. Folke, C. S. Holing and B. O. Jansson (eds) *Biodiversity Conservation: Problems and Policies*, Kluwer Academic Publishers, Dordrecht, Netherlands, pp281–300

Brookfield, H. and Padoch, C. (1994) 'Appreciating agrodiversity: a look at the dynamism and diversity of indigenous farming practices', *Environment*, vol 36, no 5, pp7–11

Brown, G. and Goldstein, J. (1984) 'A model for valuing endangered species', *Journal of Environmental Economics and Management*, vol 11, no 4, pp303–309

Brush, S. B., Taylor, J. E. and Bellon, M. (1992) 'Technology adoption and biological diversity in Andean potato agriculture', *Journal of Development Economics*, no 39, pp65–387

CBD (1992) Convention on Biological Diversity, Secretariat of the Convention on Biological Diversity, Montreal, www.cbd.int, accessed 20 July 2010

Ciriacy-Wantrup, S. (1947) 'Capital returns from soil conservation practices', *Journal of Farm Economics*, no 29, pp1181–1196

Cleveland, D. A. and Soleri, D. (eds) (2002) '*Farmers, Scientists and Plant Breeding: Integrating Knowledge and Practice*', CAB International, Wallingford, UK

Cleveland, D. A., Soleri , D. and Smith, S. E. (1994) 'Do folk crop varieties have a role in sustainable agriculture?', *BioScience*, vol 44, no 11, pp740–751

Dalton, T. J. (2003) 'A hedonic model of rice traits: economic values from farmers in West Africa', Paper presented at the 25th International Conference of Agricultural Economists, 16–22 August, Durban, South Africa

de Boef, W. S. (2000) 'Learning about institutional frameworks that support farmer management of agro-biodiversity: tales from the unpredictables', PhD thesis, Wageningen University, Wageningen, Netherlands

GON (2007) *Economic Survey 2005–06*, Ministry of Finance, Government of Nepal, Kathmandu, Nepal

Haab, T. C. and McConnell, K. E. (2002) *Valuing Environmental and Natural Resources*, Edward Elgar, Northampton

Hoyos, D. (2010) 'The state of the art of environmental valuation with discrete choice experiments', *Ecological Economics*, vol 69, no 3, pp1595–1603

Joshi, G. R. and Bauer, S. (2005) 'Analysis of the farmers' demand for rice varietal attributes in the Terai region of Nepal', Paper presented at the Deutscher Tropentag Conference, 11–13 October, Stuttgart-Hohenheim, Germany

Joshi, G. R. and Bauer, S. (2006) 'Determinants of rice variety diversity on household farms in Terai region of Nepal', Paper presented at the International Association of Agricultural Economists Conference, 12–18 August, Gold Coast, Australia

Lancaster, K. (1966) 'A new approach to consumer theory', *Journal of Political Economy*, no 74, pp132–157

Mayer, E. and Glave, M. 1999. '*Alguito para ganar* (a little something to earn): profits and losses in peasant economies', *American Ethnology*, vol 26, no 2, 344–369

MOAC (2006) 'Statistical information on Nepalese agriculture', Ministry of Agriculture and Cooperatives, Kathmandu, Nepal

Oldfield, M. (1989) '*The Value of Conserving Genetic Resources*', Sinauer Associates, Sunderland, MA

Pearce, D. and Moran, D. (1994) *The Economic Value of Biodiversity*, Earthscan, London

Perrings, C. S., Maler, K. G., Folke, C., Holling, C. S. and Jansson, B. O. (eds) (1995) *Biodiversity Loss: Economic and Ecological Issues*, Cambridge University Press, Cambridge, UK

Poudel, D. and Johnsen, F. H. (2009) 'Valuation of crop genetic resources in Kaski, Nepal: farmers' willingness to pay for rice landraces conservation', *Journal of Environmental Management*, vol 90, no 1, pp483–491

Rana, R. B., Garforth, C., Sthapit, B. and Jarvis, D. (2007) 'Influence of socio-economic and cultural factors in rice varietal diversity management on-farm in Nepal', *Agriculture and Human Values*, vol 24, no 4, pp461–472

Reid, W. V., Laird, S. A., Meyer, C. A., Gamez, R. S. A., Janzen, D. H., Gollin, M. A. and Juma, C. (1993) 'A new lease on life', in W. V. Reid, S. A. Laird, C. A. Meyer, R. Gamez, A. Sittenfeld, D. H. Janzen, M. A. Gollin and C. Juma (eds) *Biodiversity Prospecting: Using Genetic Resources for Sustainable Development*, World Resources Institute, Washington, DC

Rhoades, R. E. and Nazarea, V. D. (1999) 'Local management of biodiversity in traditional agroecosystems', in W. W. Collins and C. O. Qualset (eds) *Biodiversity in Agroecosystems*, CRC Press, Boca Raton, FL, pp215–236

Ridker, R. G. (1967) *Economic Costs of Air Pollution Studies in Measurement*, Praeger, New York, NY

Rosen, S. (1974) 'Hedonic prices and implicit markets: product differentiation in pure competition', *Journal of Political Economy*, vol 82, no 1, pp34–55

Upreti, B. R. and Upreti, Y. G. (2002) 'Factors leading to agro-biodiversity loss in developing countries: the case of Nepal', *Biodiversity and Conservation*, no 11, pp1607–1621

Vandermeer, J. M., van Noordwijk, J., Anderson, C. O. and Perfecto, I. (1998) 'Global change and multi-species agroecosystems: concepts and issues', *Agriculture, Ecosystems and Environment*, vol 67, no 1, pp1–22

Waugh, F. (1928) 'Quality factors influencing vegetable prices', *Journal of Farm Economics*, 10, no 2, pp185–196

Wood, D. and Lenne, J. M. (eds) (1999) *Agrobiodiversity: Characterisation, Utilization and Management*, CABI Publishing, Wallingford, UK

Zimmerer, K. S. (2002) 'Social and agroenvironmental variability of seed production and the potential collaborative breeding of potatoes in the Andean countries', in D. A. Cleveland and D. Soleri (eds) *Farmers, Scientists and Plant Breeding: Integrating Knowledge and Practice*, CAB International, Wallingford, UK, pp83–107

Chapter 4

Farmers' Perceptions on Replacement and Loss of Traditional Crop Varieties: Examples from Ethiopia and Implications

Edilegnaw Wale

Summary

There is widespread concern that adoption of more uniform, improved crop varieties narrows crop varietal diversity on-farm due to possible replacement and loss. If the replacement and loss are really occurring, there will have to be mitigating measures. Be that as it may, there are diverse views on the question of replacement and loss. Notwithstanding their policy relevance, farmers' concerns and perceptions on the questions of replacement and loss have not been given explicit attention in this discussion. Indeed, this has been an interesting and yet neglected area of enquiry in genetic resources policy research. Farmers' perception is the key determinant of their actions and their actions, in turn, are the key determinants of their contribution in terms of on-farm conservation. In addition to mainstreaming and enriching the essence of the discussion, investigations on farmers' perceptions can inform decision-makers.

Based on these premises, this paper explores farmers' views on replacement and loss and their importance to their livelihoods. To that end, perception data were elicited from 395 farm households in northern Ethiopia. The descriptive statistics show that, even though there is variation, the majority of the sampled farmers agree that replacement and loss are happening and this trend is decreasing the chance to find traditional varieties on their fields. The logit regression results further show that the important variables to explain farmers' perception, and what the loss means to their livelihoods, are farmers' networks, involvement in agricultural extension, chance in utilizing improved seeds, variety attribute preferences, market constraints, livestock ownership and the frequency of food shortages that farmers face. Based on the empirical results, the chapter concludes that future on-farm conservation strategies have to target farmers who better understand the occurrence of replacement and loss and those who better appreciate the role of traditional varieties to their livelihoods. Enhancing local interactions with such farmers can broaden the spill-over effects in terms of awareness.

Introduction

Crop diversity loss is one of the emerging but less visible rural development problems in Ethiopia. While landraces are used and maintained in traditional agriculture, the objective of farmers in such systems is not so much conservation but economic survival (Hardon, 1996). Farmers maintain traditional varieties only if the varieties generate private benefits, address household concerns and support their livelihoods. As noted in Chapter 1, this does not ensure maintenance of all components of crop diversity as some varieties might be marginalized and lost.

The issue of loss of crop genetic resources has been a concern since the 1930s (Harlan and Martini, 1936). The major focus of the discussion is how to address the potential conflict between the dissemination of agricultural technologies (or intensification of agriculture) and maintaining agro-biodiversity. Much of the concern is that local diverse crop varieties will be replaced by uniform improved ones. If the replacement is occurring, there will have to be mitigating measures.

Farmers' perceptions are important in influencing their crop variety choice and use (Tripp, 1996). Their perception shapes their actions in terms of variety choice which, in turn, affects on-farm conservation and crop varietal diversity. For instance, if they view local varieties as better adapted to the local (and harsh) environment (pests, diseases, erratic rainfall, etc.), they will keep on using traditional varieties to the extent that adaptability is their concern. If they appreciate the role of traditional varieties to their livelihoods (e.g. stabilizing incomes), they will maintain these varieties to the extent that they value stable income. Thus, to integrate their knowledge in future on-farm conservation policy, there is a need to understand their views and perceptions. If the perception patterns across farmers can be established, they can be used in the formulation of on-farm conservation policy acceptable to farmers contextually.

Much of the on-farm conservation policy recommendations thus far have been based on the presumption that farmers will comply with scientific recommendations. Farmers' propensity to collaborate with policy measures targeted to enhance their incentives to ensure sustainable use of crop genetic resources, however, depends on their compatibility with their perceptions and thoughts. Moreover, understanding farmers' perceptions can help identify pro-diversity farmers (or communities). It is very important to identify partner farmers (Maxted et al, 2002) and to strengthen the linkages of these farmers with others for better on-farm conservation outcomes (Subedi et al, 2003).

Studying the impact of farmers' perceptions about farm input characteristics on adoption has been receiving attention in recent times (for example, see Adesina and Baidu-Forson, 1995; Negatu and Parikh, 1999; Batz et al, 1999). However, there is little research on what their perceptions and thoughts are about traditional crop varieties and their loss. Even though farmers are custodians of the available agro-biodiversity, there is hardly any research on their perception about replacement and loss. The only study known to the author is a comparative analysis of farmers' perceptions of biodiversity from more developed economies (Estonia and Finland) (Herzon and Mikk, 2007). Accordingly, farmers of both countries rated intensification of agriculture as the major driving force behind farmland bird declines.

Based on these premises, this chapter seeks to investigate farmers' views on the problems of replacement and loss of traditional varieties. It will also look into whether or not farmers consider replacement and loss relevant to their livelihoods. To this effect, the paper is organized as follows. The next section deals with the controversies on the questions of replacement and loss, mainly drawing from the literature. It highlights the diversity of views on the issue. The third section discusses the data collection process which is followed by a brief presentation of some contextual information about the study areas. The fifth section presents a qualitative description of farmers' perceptions which, among other things, shows variation in perceptions across farmers, and this is followed by a description of the variables used in the regression meant to explain this variation. Following that, the penultimate section presents and discusses regression results. Based on the empirical results, the chapter concludes with the implications for purposive (incentive-based) on-farm conservation strategies.

Do new varieties have a crowding out effect?

Detecting and assessing genetic erosion is the first priority in any major effort to arrest loss of genetic diversity (Yifru and Hammer, 2006). According to Ehrlich and Wilson (1991), the loss of biodiversity should be of concern to everyone for three basic reasons:

1   ethical and aesthetic reasons that give moral responsibility;
2   ethnocentric/utilitarian benefits derived in the past and to be foregone in the future; and
3   an array of essential ecosystem services provided with no substitutes.

Economic (e.g. Perrings et al, 1995), legal[1] (Wolff, 2004) and agro-ecological/environmental explanations have been given for loss of traditional varieties of crops. The major conclusion of the economic literature is that the prevailing incentives and institutions do not account for the unintended negative impacts (such as crowding-out effects of introduced varieties) of economic activities (Wale et al, 2009). The diagnosis of the underlying causes of loss offered by economists has tended to focus on the role of institutions/property rights and policy in distorting the operations of agricultural commodity markets (Perrings and Lovett, 1999). Accordingly, the causes of the loss are economic (market failures), institutional (property rights) and social factors (Nunes et al, 2003). More details have been provided in Chapter 1 of this book.

Changes in agricultural activities and land-use practices affect the cultivation of certain crop varieties from the available stock (Virchow, 2003). New varieties could either add or reduce diversity in lieu of replacements of the local materials. Adoption of improved varieties can result in the replacement of one type of germplasm for another and/or the integration of new genetic materials into existing gene pools through gene flows. While the first is a human-driven process, the second is governed by the natural process of gene flow and integration (Cooper et al, 2005). As noted below, the literature suggests a considerable lack of consensus on the issue of replacement and traditional variety loss.

In general, there are two polarized views. While some authors take the occurrence of replacement for granted and blame the Green Revolution for most of the loss of agro-biodiversity (Tilman, 1998; Matson et al, 1997), others question this and find implicit and unfounded assumptions to those claims (Smale, 1997; Wood and Lenne, 1997).

The first view, which can be named as the genetic erosion model, is the one that assumes that intensification of agriculture is leading to replacement.[2] Genetic uniformity has been a concern for development practitioners during the last four decades. It is motivated by the potential replacement of traditional varieties by the improved and uniform ones (Brush, 1991). Accordingly, replacement is happening and threatening the diversity of the respective crops (Cleveland et al, 1994; Tripp, 1996; and Bardsley and Thomas, 2005). Modern cultivation practices exert a direct impact on species richness (Potvin et al, 2005). Changes in the biophysical (climate change, desertification, land degradation, etc.), socio-cultural environment (education, culture) and emerging opportunities (new Green Revolution technologies) are further triggering the biophysical change (Cleveland et al, 1994), leading to a reduction in landrace diversity (Bardsley, 2003), and resulting in biophysical and socio-economic costs in many parts of the world, in both the global North and South (Thrupp, 2000).

The second view, which can be named as the complementarity school of thought, takes improved varieties as having an important role in maintaining and even enriching genetic diversity of the local gene pool and indigenous cropping systems (Dennis, 1987; Almekinders et al, 1994; Wood and Lenne, 1997). Accordingly, there is a useful complementarity between the use of improved and traditional varieties.

The claim that the Green Revolution caused genetic erosion cannot be established against a counterfactual because of the difficulties in measuring genetic erosion and proving causality in the presence of multiple intervening factors (Smale, 1997). There are unfounded and implicit assumptions within the claim (Perales et al, 1998):

- modern varieties are always superior to traditional varieties in economic profitability; and
- net return or profit maximization is the only motivating factor in farmers' variety choice decisions.

Regarding the first implicit assumption, results of some studies suggest that improved varieties may not always have a clear advantage over local varieties in terms of income (Perales et al, 1998; Wale, 2006) and there is no need necessarily to adopt improved varieties to improve productivity (Brush et al, 1988). On the second point, there is a whole body of literature in development economics that shows the diversity and complexity of objectives of small farmers. Accordingly, it is too simplistic to equate variety choice and replacement with net return alone (Brennan and Byerlee, 1991; Bellon and Taylor, 1993; Smale et al, 1994).

Moreover, according to the partial technology adoption literature (Just and Zilberman, 1983), farmers do not adopt a new technology and forget old ones automatically for safety-first behaviour and learning reasons (Smale et al, 1994). Some studies have also shown that, for satisficing reasons (like capturing desirable consumption and production traits), farmers who adopt improved varieties often continue to plant local varieties (Smale and Heisey, 1995; Brush, 1995) and complete displacement may not occur (Brush et al, 1992).

Presenting the various views above should not imply an argument against improved seeds. The institution of on-farm conservation does not reject modern agriculture or its associated technologies (Bardsley and Thomas, 2005). Rather, it presents the case that strengthening farmers' capacity to conserve agro-biodiversity includes addressing the poverty with which that diversity is associated (Tripp, 1996). Even though highest diversity is correlated with traditional farming (Perfecto et al, 1996), rural people cannot be locked into their unproductive agricultural practices (Srivastava et al, 1996) and smallholder agriculture cannot be a museum of repository for traditional varieties (Bardsley and Thomas, 2005).

Overall, there seems to be a reasonable agreement on the absence of automatic displacement, the inevitability of the change in population structure,[3] and the cumulative and slow nature of the replacement process.[4] There are no clear-cut reasons for farmers to automatically displace traditional varieties because all crop varieties have both desirable and undesirable attributes, and no crop variety (even a combination of varieties) has all the diverse traits that farmers demand. One of the reasons for lack of adequate commitment in conservation of agro-diversity is that the loss is not automatic and its impacts are gradual. The impacts are not vividly felt until it is too late and only a serious crisis will provoke society into meaningful action (Thomas Eisner, as quoted by Powledge, 1998).

Given the diversity of views, this chapter is meant to add farmers' perspectives to this discussion and pave the way for integrating their thoughts in future on-farm conservation policy.

Sampling and data collection procedures

The field work for this research was undertaken in the northeastern part of Ethiopia. Discussions were held with a total of 73 key informants (10–14 per session) during the 2006–2007 cropping season. Moreover, information was also collected from concerned individuals and records of local government agricultural offices.

The study considered two zones (South Gondar and North Wollo) in Amhara Regional State.[5] In South Gondar, Farta and East Estie districts (called 'Woredas') were considered whereas in North Wollo, Guba Lafto Woreda was considered. In each Woreda, two peasant associations (PAs) were selected, i.e. Mainet and Kolley Dengorse in Farta, Shimagle Giorgis and Sholekt in East Estie and Ala Wuha and Woinye in Guba Lafto.

The villages surveyed in Mainet include Eyesus, Antira, Guasat and Mainet. Villages called Dengorse, Kona, Gundiba, Libanos, Egziabher-Ab and Kolley were surveyed in Kolley Dengorse. In Shimagle Giorgis, Giorgis, Tirtiriat, Medhanie-Alem and Mehalgie are the surveyed villages. For Sholekt, the villages include Modir, Mehalgie, Enzona and Sholekt. In Woinye, Shall, Mamuarecha, Kokeba and Kolegenda were surveyed while the villages surveyed in Ala Wuha were Doro Gibir, Hayqote, Ambalat, Guasat, Kulu Baine and Woineba.

In consultation with agricultural bureaus and development agents (DAs) working with farmers, the survey was structured to cover a wide range of villages and farmers. To this effect, finding the most important sources of heterogeneity of peasant households was the most important step. By and large, poverty status/wealth and accessibility (roads and markets) were the two most outstanding sources of heterogeneity. In the sampling exercise, farm households of each poverty group (using wealth ranking done by the key informants) and those found in sites close to markets/roads and far away from markets/roads were considered. The other variables considered during the sampling (in the respective PAs) include importance of income sources outside agriculture, education and gender. The records from local agricultural offices were used to sample the farm households in the respective villages. In each PA, five to ten reserve farmers were also selected in case the sampled farmers went missing.

In terms of PA distribution, 76, 68, 68, 64, 64 and 55 farm households were sampled from Shimagle Giorgis, Ala Wuha, Mainet, Sholekt, Woinye and Koley Dengorse, respectively. In total, the sample involves 2 zones, 3 Woredas, 6 PAs, 25 villages and 395 farm households. Essentially, multistage stratified random sampling was the strategy used.

The draft questionnaire was then pre-tested to check its feasibility in line with farmers' expectations, responses and understanding. About 18 farmers were involved during the pre-test. Following the pre-test, the necessary changes were made to the draft questionnaire to ensure the consistency and validity of each question. The information from the key informant interviews was also used to refine the household survey instrument. In total, 26 enumerators who knew the local language and culture were hired and trained to do the personal interviews.

Moreover, 3 facilitators, who were mainly engaged in making appointments and locating farmers, were involved.

All of the enumerators, employees of the Woreda Agriculture Bureau as DAs, were diploma holders with extensive experience in assisting farmers in the villages. Utilizing the local DAs was helpful to internalize the meaning of what farmers say. Some farmers were not able to identify varieties by names while other farmers gave different names for the same variety. Whenever such difficulties arise, the DAs were consulted to verify variety types and sort out differences, similarities and inconsistencies.

The sampled farmers were interviewed from mid-October 2006 to mid-January 2007. The data were then entered using SPSS (version 14) by professional data-entry clerks. Having coded all the open-ended questions and entered all the valid data, it was decided to go to the villages for the second time to rectify some of the key variables where the responses were found to be invalid/inconsistent. A total of 123 questionnaires were revised in the field during the last three weeks of April 2007.

Some contextual information about the study areas

Agro-ecologically, East Estie Woreda has Woina dega (midland, 1500–2500m above sea level) accounting for 66 per cent of the area of the Woreda, Dega (highland, 2500–4000m above sea level) accounting for 27 per cent and Kola (low land, 1400–1500m above sea level) accounting for 6 per cent. The average annual rainfall (RF) in this Woreda ranges from 1300 to 1500mm. The temperature ranges from 8 to 25°C. The average temperature in Farta Woreda ranges from 9 to 25°C and the average annual RF is 900 to 1099mm per annum. This Woreda is situated 1500 to 4135m above sea level. It is found 11°–32° to 12°–03° Eastern longitude.

Guba Lafto, a rural Woreda, is 0 to 50km from Woldeya, the capital city of North Wollo zone. Agro-ecologically, most of the PAs are Woina dega with the exception of Ala Wuha which is mainly Kola (95 per cent). In terms of altitude, the PAs range from 500 to 3000m above sea level. The average temperature is about 25°C with annual average RF of 1000 to 1500mm. The soil types are by and large loam and land degradation is one of the cross-cutting problems in all PAs.

There is one typical PA worth mentioning, namely Woinye. The landscape in this PA is typically rugged and mountainous. It is located 23km west of Woldeya town. Irrigation is more common in this PA, which has enabled about 25 per cent of the farmers to produce during both 'meher' and 'belg' seasons. Cash crops like vegetables, fruits and sugarcane are very common in this PA.

In each PA, primary and secondary (mostly up to eighth grade) schools are available. Most of the PAs also have access to all-weather roads while all have seasonal/dry weather roads. There is also telephone communications in each PA.

There is at least one safe (treated) drinking water well in each PA. It takes farmers 30 to 120 minutes walk to reach markets.

Coming to an overview of the sampled farmers, about 88 per cent of the heads of the sampled farm households are male, and 54 per cent of the heads of the sampled households cannot read and write. Only 26 per cent of the heads of the households have received a level of education better than reading and writing through church, mosque and/or schools. For 91 per cent of the farm households, agriculture is the most important economic activity. For the rest, household activities (mainly for female-headed households), non-farm small-scale economic activities and working in the church are the most important activities supporting the household. More than half (53 per cent) of the respondents are net buyers of agricultural products.

Farmers' views on replacement and loss: insights from the survey

Given the diversity of views presented in the section titled 'Do new varieties have a crowding out effect?' above, we now discuss the farmers' perspectives. To generate data, questions were framed and presented to farmers one by one to elicit their views on replacement and loss, and what this means to their livelihoods. Following their agreement or disagreement, they were further requested to give reason(s) for their response. Their reasons are coded and presented in the tables that follow.

To start with, if farmers do not see a change in the composition of the varieties they grow over the course of many years, they will not be able to appreciate the occurrence of replacement and loss. Hence, the first question was 'How far do you agree with the idea that the composition of the crop varieties in your village has changed in the last five years?'[6] The data show that most farmers (88 per cent of the 373 respondents) agree. Table 4.1 below reports their reasons.

Table 4.1 *Farmers' thoughts on the change in the composition of varieties*

Reasons of those who think that the composition of varieties in their villages has changed	% of respondents (n=373)
In most of the cases, the local varieties have been replaced by the improved ones	24
The productivity of traditional varieties is decreasing every year	21
There is frequent supply of various seeds resulting in the change	14.4
The improved seeds themselves will be changed to local ones as time goes on	10
The local varieties are getting old	8.5
Over time, varieties have been susceptible to frost and wind	8
The types of varieties (colour, size, yield, etc.) we see differ from one year to another	4.2
Others	9

Source: 2007 GRPI survey data, Ethiopia

Table 4.2 *Farmers' views on the difficulty of finding traditional varieties of crops in their villages*

Reasons of those who believe that it is getting hard to find traditional varieties of crops in the villages	% of respondents (n=393)
Since we are using the new seeds, the old ones are getting lost	35.6
The local varieties are not productive; their yield is dwindling every year	22.9
The traditional varieties are no longer in the hands of many farmers	6.6
The old varieties are no longer tolerant to drought	4.6
If the local seeds are used for many years, they become susceptible to diseases	2.3
Traditional varieties are no longer compatible with the type of soil we are dealing with	1.5

Source: 2007 GRPI survey data, Ethiopia

According to Table 4.1, replacement of traditional varieties by the improved ones and frequent supply of improved seeds are important reasons for farmers to think that there is a change in the portfolio of varieties on-farm. This is in line with the notion that increasing the rate of varietal improvement speeds varietal replacement (Heisey and Brennan, 1991). It is striking to notice that some farmers are knowledgeable about the adaptation of improved varieties to local ones due to the evolution that occurs in response to agro-ecological dynamics.[7]

About 12 per cent of the respondents do not see a change in the composition of the crop varieties they grow in their village. This is probably because they are typically dependent on traditional varieties of crops. For some of them, there are no improved seeds at all: only the land is changing (for the worse), not the seeds.

The second question was 'How far do you agree with the idea that it is becoming ever difficult to find traditional varieties of crops in the village and on farmers' fields?' About 74 per cent of the 393 respondents agree with this idea. Table 4.2 reports their reasons.

According to the preceding table, farmers think that it is becoming harder to find traditional varieties of crops due to their replacement by the new ones as the traditional varieties are becoming less productive, no longer tolerant to drought, susceptible to diseases and incompatible with the type of soil farmers are dealing with. For livestock, Zander et al (2009) have shown that local livestock keepers (about 34 per cent in Ethiopia and 21 per cent in Kenya) agreed that it is increasingly becoming harder to find pure Borana cattle in local markets.

Farmers understand the general pattern of yield deterioration in their own varieties (Heisey and Brennan, 1991) and make a replacement decision accordingly. Apparently, farmers who agree with the preceding question are getting better access to improved seeds and their preferences are for high yield and tolerance to environmental stress (drought and diseases).

About 26 per cent of the respondents feel that it is not difficult to find traditional varieties of crops in the hands of most farmers. Their reasons are that old varieties are still in the hands of many farmers and/or old varieties are easily adaptable to local harsh weather conditions.

Table 4.3 *Farmers' views on the inevitability of growing traditional varieties of crops*

Reasons of those who agree	% of respondents	Reasons of those who do not agree	% of respondents
The local seeds are well adapted to the harsh local environment	48	Due to susceptibility to drought, frost and chill, the local varieties are no longer useful	17.6
The local varieties can still give some yield however bad the situation is	24		
Local varieties are tolerant to drought/frost	11.7		
It is better to use our own seeds than getting exposed to unforeseen problems	4.7	The land is no longer productive with local varieties, yield is deteriorating	4.7
In most of the cases, the yield from local varieties is better	2.1		
There simply is no other choice	1.8	The new varieties are better and tolerant to drought	3.1
I have never used improved seeds	1		
We cannot afford to buy the new varieties	0.3	As long as the new varieties are there, I do not worry about the local ones	2.9

Note: The total sum exceeds 100 because some farmers (who are nested) have given multiple reasons and the reasons have been grouped for reporting purposes; n=384.
Source: 2007 GRPI survey data, Ethiopia

The third question was 'How far do you agree with the idea that growing traditional varieties of crops is inevitable for better adaptation to harsh local conditions?' About 76 per cent of the 384 respondents agree while the rest think otherwise. Table 4.3 summarizes their respective reasons.

Most farmers feel that growing traditional varieties of crops is inevitable because these varieties are better adapted to harsh local conditions and can still give some yield however bad the situation is. Landraces often tend to be low yielding but dependable and adapted to crude cultural practices and poor soil quality (Harlan, 1975). For others, traditional varieties are better known to them and do not carry the risk of becoming exposed to an unforeseen situation. For some farmers, there is simply no other choice, and they lack the experience and capacity to take on the new varieties.

For farmers who do not believe that growing traditional varieties of crops is inevitable for better adaptation to harsh local conditions, their reasons had to do with failure of the varieties to be productive enough, lack of tolerance to local stress and poor quality of their land for growing local varieties.

Table 4.4 *Farmers' thoughts on replacement and loss*

Reasons of those who think replacement and loss is happening	% of respondents	Reasons of those who do not agree	% of respondents
We are happy with the yield level of the improved seeds and hence no longer using the local seeds	30.2	The local varieties are still in the hands of most farmers	10.9
We have abandoned the old varieties while using the new ones	29.6	We are used to the local varieties; we will still keep on using them	9.6
Local varieties are being lost gradually	7.8	The local seeds maintain the quality of land	6.2
The improved varieties are the ones dominating on-farm	6.7	It is better to face more predictable outcome than getting indebted to buy improved seeds on credit and face unpredictable outcome	5.7
The land is getting fed-up with the local seeds; the improved varieties are better suited for the land	6.4		
We are not happy with the yield level of the local varieties	2.8	The local seeds are most adapted and preferred by most farmers	4.2

Note: The total sum exceeds 100 because some farmers (who are nested) have given multiple reasons and the reasons have been grouped for reporting purposes; n=384.
Source: 2007 GRPI survey data, Ethiopia

The fourth question was 'How far do you agree with the idea that replacement of local varieties by the improved ones is happening and leading to disappearance of traditional varieties?' About 70 per cent of the 384 respondents agree while the rest do not. Table 4.4 summarizes their respective reasons.

Most farmers think that the replacement of traditional varieties of crops by the improved ones is happening and it is leading to the disappearance of the former for yield differential and land suitability reasons. According to Heisey and Brennan (1991), farmers' varietal replacement decision can be the result of yield deterioration. The land seems to be no more in harmony with the demands of local seeds. In some cases, the improved varieties are found to be more suited to the land and the potentials of the traditional varieties are exhausted. Farmers often note that varieties become 'tired' and need replacement (Wood and Lenne, 1997). In sum, the extent to which farmers adopt improved varieties and replace their landraces depends on the extent to which the new varieties better satisfy their household livelihood strategies (Maxted et al, 2002).

Farmers' reasons for thinking that replacement and loss is not happening are based on their trust in the local varieties in terms of yield, better adaptation and

Table 4.5 *Farmers' thoughts on the effect of loss of traditional varieties of crops on their livelihoods*

Reasons of those who think they are affected	% of respondents	Reasons of those who do not think they are affected	% of respondents
There is no improved seed without the local varieties	30	As we are getting the improved seeds (with good traits) in adequate quantities, we are not affected by the absence of local varieties	22
Improved seeds are so expensive, unlike the local seeds which are accessible mostly from own sources	24		
The local ones are good for food	9.8	As local varieties are not high yielding, we do not miss them	6.2
The local varieties are good for health	6.5		
There are so many useful traditional seeds and losing them is so painful	4.1	Improved varieties are resistant to drought	2

Note: The total sum exceeds 100 because some farmers (who are nested) have given multiple reasons and the reasons have been grouped for reporting purposes; n=369.
Source: 2007 GRPI survey data, Ethiopia

maintaining soil quality. Moreover, farmers' better knowledge about local varieties and their intention to avoid indebtedness are also important factors.

The last question was 'Do you believe that loss of traditional varieties of crops is influencing your livelihoods?' About 71 per cent of the 369 respondents have responded affirmatively while the rest do not believe so. Their respective reasons are contained in Table 4.5.

Most farmers are aware of the impact of the loss of traditional varieties of crops on their livelihoods in terms of input cost, income, health and nutrition. Accordingly, local varieties are not only sources of improved seeds but also they are easily and cheaply accessible. Farmers' knowledge about the utility of traditional varieties is thorough and instructive.

On the contrary, farmers who do not think that their livelihoods are affected by loss of traditional varieties have a lot of trust in the improved seeds. These farmers, unlike the previous ones, seem to think that improved seeds are able to solve most of their agricultural problems. Accordingly, improved seeds are perceived to be superior.

In summary, the discussions held with farmers suggest that yield is the overriding factor in farmers' variety choice decisions. Land degradation and deterioration of the productive potential of farmers' varieties is leading to the loss of traditional varieties of crops because some of farmers' varieties are retiring (losing their desirable attributes). Drought and desertification are the ever-increasing problems faced by many farmers that are influencing variety choice. Hence, measures that enhance the productivity of farmers' varieties (e.g. irrigation, land husbandry practices, integrated pest management, etc.) can support on-farm conservation (Heisey and Brennan, 1991).

Now that farmers' views have been described, the next question is 'Why do farmers differ in their views concerning variety replacement, loss and the impact of such an event on their livelihoods?' The rest of the paper will dwell on the possible reasons, considering the responses as dichotomous variables.

Description of variables used in the logit analysis

Farmers perceive the performance (with respect to different traits) and utility of local and improved varieties differently. Their perception regarding loss of crop diversity and its impact on their livelihoods can be shaped by a multitude of factors. Essentially, farmers' heterogeneity with respect to those factors has to explain their differences in perception. For instance, two of those factors include the suitability of their working environment and resource endowment.

For this study, variables that can capture household contextual characteristics (age, gender and education level of the household head), social networks, poverty/wealth status, experience in growing improved seeds/agricultural extension, access to markets/roads/improved seeds, variety attribute preferences and own experience in abandoning local varieties are the ones considered to explain perceptions. These variables are identified based on theory, previous literature, the descriptive statistics in the previous section and field observations during the discussions held with farmers.

Subjective perceptions of farmers about technologies and consumers of agricultural technologies influence their adoption decisions (Adesina and Baidu-Forson, 1995). Similarly, one can argue that subjective perceptions of farmers about loss of traditional varieties and the impact of such a loss on their livelihoods (as acknowledged by farmers) influence their variety choice and use decisions. That is the motivation to study farmers' perceptions.

The first variable considered is farmers' age because perception is the cumulative knowledge of farmers' experience. The older farmers are, the higher the chance for them to notice the occurrence (if any) of replacement and loss, and the higher the chance for them to notice that their livelihoods have been impacted.

Education involves exposure to new ideas and knowledge that affects farmers' perceptions. In most of the cases, better educated farmers are better informed, and they are ready to take and exercise new technologies. Therefore, better educated farmers can be expected to perceive that loss and replacement is occurring. Similarly, farmers' networks affect perceptions because information exchange among farmers, which occurs through their networks, makes them aware, i.e. more networked farmers are more likely to perceive replacement and loss happening.

Women farmers have typically great preference for varieties with food and cooking quality attributes (Bellon and Risopoulos, 2001). However, since it is not *a priori* known as to which varieties (local or improved) are superior in terms of food-quality traits and since they are crop-specific, the impact of gender on perception remains unpredictable.

If crop production is the most important income source for the household, the farmer is more likely to notice changes in variety type and composition. Because of their engagement with crop production, farmers are also more likely to miss the local public goods value[8] of traditional varieties. Similarly, net sellers of agricultural products are more likely to perceive changes in the crop variety portfolio, replacement and loss. Their livelihoods will also be more influenced by the loss.

A farmer's perception may be shaped by their experience of growing the new variety, extension services, and their knowledge about the modern variety (Negatu and Parikh, 1999). Most farmers who have been taking part in the agricultural extension programme and/or have longer experience with improved varieties can be expected to notice the possibility of replacement and loss. Those farmers who have got a better chance to get improved seeds (due to access or better purchasing power) can also be expected to perceive the occurrence of replacement and loss in their respective villages.

Poor farmers who often face food shortages often depend on traditional varieties. Consequently, they do not believe that replacement and loss is happening. On the contrary, those better-off farmers who cultivate large farms and own more livestock are mostly able to take risk and try new genetic technologies. These farmers are, therefore, more likely to see the replacement and loss process happening.

Farmers' variety attribute preferences are the other important factors that can influence and shape their perceptions. Subsistence farmers who have more preference for yield stability and early maturity will have a higher chance of growing traditional varieties of crops if these traits are better embedded in traditional varieties of crops. Those farmers with greater preference for higher price and yield are more likely to be using mainly improved seeds if these attributes are better embedded in improved varieties. Since there was no *a priori* information on which attributes are embedded in which varieties (improved or local), the relationship could not be predicted.

Perceptions, by and large, are derived from the individual's own behaviour and actions. Farmers' own experience in abandoning local varieties is, therefore, another important variable which is expected to shape their perceptions. Those farmers who have abandoned one or more local variety of a crop are more likely to notice the change in the composition and perceive occurrence of replacement and loss. Tables 4.6 and 4.7 describe the complete list of variables and present the above hypothesized relationships.

Table 4.6 *Description of variables to explain farmers' perception towards the replacement of local varieties of crops by improved ones and expected signs*

Variable	Description	Mean (SD)	Expected sign
Percept2	Do you agree with: 'Replacement of local varieties by the improved ones is happening and leading to disappearance of traditional varieties'? (1 = Yes, 0 = No)	0.70 (0.47)	Dependent variable
Age	Age of the household head (years)	48.8 (14.0)	+
Education	The highest level of schooling completed by the household head (grade)	1.28 (2.31)	+
Inc_sorc	Is crop production the most important income source for the household? (1 = Yes, 0 = No)	0.87 (0.34)	+
Soc_net	The number of local networks that household members are engaged in (such as cooperatives, labour sharing, local office, church, and credit/saving institutions)	5.40 (2.45)	+
Livestock	The natural logarithm of total value (in Birr) of livestock owned	8.10 (1.32)	±
Land	Land holding (in hectares) per household size	0.22 (0.14)	±
Extension	Have you ever participated in the agricultural extension package program? (1 = Yes, 0 = No)	0.67 (0.47)	+
YearsIV	Farmers' experience with improved seeds (years)	3.73 (3.94)	+
Dist_road	How much time do you need to travel on foot to reach the nearest dry weather road (in minutes)?	45.64 (47.3)	−
Chance	Do you have a chance to buy or get improved seeds whenever you want? (1 = Yes, 0 = No)	0.65 (0.48)	+
Sel_buy	In most of the years, do you sell more (in monetary terms) agricultural products than you buy? (1 if net seller and 0 if subsistence and net buyer)	0.37 (0.48)	+
EM_rank	In comparison to other traits listed, how do you rank a variety which matures early?	4.36 (2.20)	±
P_rank	In comparison to other traits listed, how do you rank a variety which fetches the highest price?	3.82 (1.90)	±
No_Aband	Number of traditional varieties of crops abandoned by farmers the last ten years	1.72 (1.29)	+

Source: 2007 GRPI survey data, Ethiopia

Data analysis, results and discussions

Farmers' perceptions drive their decisions to maintain traditional varieties of crops. Thus, explaining perceptions is an input to establish the targeting mechanisms for on-farm conservation, to test the policy relevance of scientific recommendations and to harmonize development interventions with crop

Table 4.7 *Description of variables to explain farmers' views on the importance of maintaining local varieties to their livelihoods*

Variable	Description	Mean (SD)	Expected sign
Percept3	Do you agree with: 'Loss of traditional varieties of crops is influencing our livelihoods'? (1 = Yes, 0 = No)	0.71 (0.46)	Dependent variable
Age	See Table 4.6		+
Education	See Table 4.6		+
Gender	Gender of the head of the household (1 = Male; 0 = Female)	0.88 (0.33)	±
Inc_sorc	See Table 4.6		+
Soc_net	See Table 4.6		+
Livestock	See Table 4.6		±
Fod_short	The number of months in a year that the household has problem of food shortage (on average)	2.17 (2.81)	+
Extension	See Table 4.6		−
YearsIV	See Table 4.6		−
Chance	See Table 4.6		−
Mkt_rank	See Table 4.6		+
YS_rank	See Table 4.6		±
Fod_rank	Farmers' preference for a variety which is good for food (taste, colour, milling quality, etc.)	4.32 (2.19)	±
Y_rank	Farmers' preference for a variety that is better yielding	2.37 (1.56)	±
FV_aband	See Table 4.6		−

Source: 2007 GRPI survey data, Ethiopia

diversity outcomes. To gain more insights on these issues, from among the questions presented to farmers and discussed in the previous section, farmers' perceptions regarding questions 4 and 5 are further analysed using the logit model which takes the form:

$$Y = \frac{\exp(z)}{[1 + \exp(z)]}$$

(Eqn 4.1)

where z refers to the explanatory variables discussed and hypothesized in the previous section and presented in Tables 4.6 and 4.7. The regression results (Tables 4.8 and 4.9) and related discussions are presented subsequently.

Overall, the model rightly (see Table 4.8) predicts 75.2 per cent of the farmers who think that local varieties are being replaced by improved ones and 59.5 per cent of those who think otherwise. The overall prediction rate is 73.2 per cent. The

Table 4.8 *Logistic regression results to explain farmers' perception towards the replacement of traditional varieties of crops by improved ones*

Variable	Coefficient (z)	Marginal effects: dy/dx
Age	−0.0075 (−0.77)	
Education	0.051 (0.78)	
Inc_sorc	−1.37 (−2.66)***	−0.197
Soc_net	0.211 (3.46)***	0.040
Livestock	−0.177 (−1.43)	
Land	0.602 (0.69)	
Extension	−0.491 (−1.55)*	−0.09
YearsIV	−0.02 (−0.55)	
Dist_road	0.005 (1.46)	
Chance	1.051 (3.58)***	0.213
Sel_buy	−0.817 (−2.97)***	−0.162
EM_rank	0.074 (1.27)	
P_rank	−0.18 (−2.42)***	−0.035
No_Aband	0.14 (1.32)	
Constant	2.69 (2.15)	
Dependent variable is Percept2	Prob > χ^2	= 0.00
Number of observations = 332	Log pseudo-likelihood	= −176.9
Wald $\chi2(14)$ = 42.4	Pseudo R^2	= 0.118

Notes: *** significant at 1% level; ** significant at 5% level; * significant at 10% level.
Source: GRPI, Ethiopian survey, 2006/2007

likelihood ratio test results (LR $\chi^2(2)$ = 49.5; Prob > χ^2 = 0.00) also show the significance of the model.

Those farmers who can easily access improved seeds seem to be more convinced that replacement and loss is happening. If they themselves can easily get improved seeds, they are more likely to consider that improved seed distribution is leading to replacement and loss.

Social networks increase farmers' awareness on what is happening in terms of the replacement of traditional varieties of crops by the improved ones, which is supported by the results. If farmers are concerned with this trend, their awareness (through the information exchange via their networks) will shape their subsequent actions and influence their engagement with on-farm conservation initiatives.

Farmers who have a greater preference for varieties which can fetch a better price are also more likely to see replacement and loss happening. This cannot be surprising as these market-oriented farmers are more likely to be using improved varieties. The results also show that net sellers of agricultural products are less likely to notice replacement and loss.

The results suggest that farmers who have taken part in agricultural extension (at least once) do not see replacement happening. Though this is against our

Table 4.9 *Logistic regression results to explain farmers' views on the importance of losing traditional varieties to their livelihoods*

Variable	Coefficient (z)	Marginal effects: dy/dx
Age	−0.0013 (−0.13)	
Education	−0.020 (−0.33)	
Gender	−0.626 (−1.25)	
Inc_sorc	0.255 (0.65)	
Soc_net	0.17*** (2.42)	0.032
Livestock	0.26** (2.35)	0.050
Fod_short	0.095** (1.94)	0.018
Extension	−0.53* (−1.74)	−0.095
YearsIV	−0.066* (−1.76)	−0.013
Chance	−0.303 (−1.04)	
Mkt_rank	−0.441*** (−3.08)	−0.084
YS_rank	−0.137** (−2.38)	−0.026
Fod_rank	0.117** (1.99)	0.022
Y_rank	−0.042 (−0.48)	
FV_aband	0.223 (0.77)	
Constant	0.396 (0.36)	
Dependent variable is Percept3	Prob > χ^2	= 0.0087
Number of observations = 341	Log pseudo-likelihood	= −184.24
Wald χ^2(15) = 31.05	Pseudo R^2	= 0.09

Notes: *** significant at 1% level; ** significant at 5% level; * significant at 10% level.
Source: GRPI, Ethiopian survey, 2006/2007

prediction, involvement in agricultural extension does not necessarily mean use of improved varieties. Moreover, the message that participating farmers get from agricultural extension officers is mainly about increasing production and productivity. It hardly raises issues of losing traditional varieties of crops. This might have influenced their attitude. Unlike our expectation, farmers whose most important income source is crop production do not seem to see replacement and loss happening, which could be due to the greater dependence of these farmers on traditional varieties.

Overall, the model rightly (see Table 4.9) predicts 75 per cent of the farmers who think that local varieties are important to their livelihoods and 66 per cent of those who think otherwise. The overall prediction rate is 74.5 per cent. The likelihood ratio test results (LR χ^2(2) = 43.2; Prob > χ^2 = 0.00) also show the significance of the model.

According to the results in Table 4.9, farmers' networks increase the chance that they value the importance of losing traditional varieties to their livelihoods. The information exchange is enhancing their appreciation of the role of traditional varieties to their livelihoods. This reinforces the role of these local networks in future on-farm conservation endeavours.

Total value of livestock owned also increases the chance that farmers will value the importance of the loss of traditional varieties to their livelihoods. This can be attributed to the production synergy between crop and livestock production because in most cases the traditional varieties of crops are preferred for livestock feed.

Farm households who have been facing more frequent food security problems appreciate the effect of losing traditional varieties of crops to their livelihoods. This result reinforces the typical role of traditional varieties of crops to the poor (Wale, 2004). Voluntarily and/or involuntarily, the poor farmers mainly grow traditional varieties and they better appreciate the desirable features of these varieties (yield stability, early maturity, environmental adaptability, etc.).

Farmers with greater preference for yield stability trait are more likely to value the loss of traditional varieties to their livelihoods. This result suggests that this trait is one of the desirable attributes of traditional varieties and hence those who have mentioned it as their priority will have to be concerned with this trend.

Those farmers who have greater preference for a variety that is good for food do not see the importance of losing traditional varieties of crops to their livelihoods. This can be because either the improved varieties that farmers are knowledgeable about are found to be good for their food traits or the local varieties have been perceived to have lost their desirable food-quality attributes. As a result, farmers are not missing food taste-related values of traditional varieties.

As expected, those farmers who are more experienced with improved varieties do not see the importance of losing traditional varieties to their livelihoods. Those farmers who have taken part in the agricultural extension programme (at least once) also do not feel the impact of losing traditional varieties to their livelihoods. Their engagement with the programme and the chance they had to use introduced varieties and inputs might have convinced them that it is possible to do crop farming without necessarily using traditional varieties. As noted above, the message that agricultural extension participant farmers often get is mainly about enhancing agricultural production and productivity, not the utility of traditional varieties of crops to their livelihoods.

Those farmers who have reported that they have been constrained by market and price-related problems, or those farmers who are engaged in commercial activities but are market-constrained, value the importance of losing traditional varieties of crops to their livelihoods. Even though this is not as expected, it can be attributed to desirable attributes of traditional varieties to consumers.

Conclusions

This study has been motivated by the idea that farmers' perceptions on the questions of traditional crop varieties replacement and loss, and what this means to their livelihoods, are important determinants of their variety choice and use decisions. However, they are hardly studied despite the importance of integrating farmers' perceptions into the design of on-farm conservation policy. The important value that this study adds, therefore, is in terms of integrating farmers' perspectives to the replacement and loss question. This, among other things, can ensure the design of policies and strategies that can enhance farmers' compliance with on-farm conservation incentives.

A partial review of the available literature shows that there is considerable controversy on the issue of replacement and loss. While some question the occurrence of loss and replacement, others take them for granted. While some consider improved varieties as threats to traditional varieties, others take them as complementary. Overall, most studies agree on the absence of automatic displacement, the inevitability of the change in crop varietal population structure, and the cumulative and slow nature of the replacement process. Thus, there is a need to carefully monitor the nature of the change before jumping into any conclusion.

Most of the on-farm conservation policy recommendations so far seem to take farmers' compliance for granted. Their compliance, however, will depend, among other things, on its compatibility with their perceptions and thoughts. It will, therefore, be imperative to understand farmers' views on the questions of loss and replacement and what it means to their livelihoods.

It is remarkable to learn that a sizable number of farmers are aware of the fact that the local varieties are the inputs/raw materials for developing the improved ones. Farmers are also knowledgeable about the process of change from improved to local varieties due to the adaptation in response to agro-ecological dynamics.

It is the view of the majority that replacement of traditional varieties of crops is happening and it is leading to the disappearance of traditional varieties, mainly for yield differential and land-suitability reasons. According to the regression results, these are farmers with better local networks, with easy access to improved seeds, and greater preference for varieties which can fetch better prices. Farmers' belief that replacement is not happening is based on their trust in the desirable attributes of local varieties. The regression results have shown that these are net sellers of agricultural products whose most important income source is crop production.

According to most farmers, loss of the traditional varieties of crops is already influencing their livelihoods in various ways. For one thing, cost of production increases as improved seeds are not physically and financially accessible to most smallholders. Secondly, these farmers are missing the health and nutrition benefits of traditional varieties of crops. The regression results have identified these farmers as those with better networks, owning more livestock, facing frequent food security problems and having greater preference for yield stability trait. On the contrary, farmers who do not think that their livelihoods are affected by the loss of traditional varieties have a lot of trust in the superiority of improved seeds in terms of solving most of their agricultural problems. The regression results identify these farmers as those who have taken part in the agricultural extension programme and those more experienced with improved varieties.

Implications for policy

The results of the study have implications as inputs for future on-farm conservation policy in various ways. First of all, there are genuine reasons to be concerned about replacement and loss, which have been confirmed by the overwhelming majority of farmers, i.e. whatever time it takes, it is occurring and mitigation measures are indispensable. Since the results suggest trade-offs between access to improved seeds and the survival of farmers' varieties on their fields, it is high time to link dissemination of improved varieties with on-farm conservation initiatives.

In terms of policy, it would be unproductive to rule out either the products of modern breeding or the products of farmers' indigenous knowledge-based breeding (Tripp, 1996). The daunting task ahead is to formulate strategies that can achieve agricultural development and agro-biodiversity conservation. Since these issues are not sequential, decision-makers will have to consider both aspects of sustainable development in the design of projects, programmes and policies.

Improved varieties should not be considered as a panacea to farmers' problems. The agricultural extension programme should focus not only on enhancing productivity through improved seeds but also on ensuring the survival of traditional varieties. To this effect, the programme has to make on-farm conservation an integral component. According to this study, farmers who have greater preference for survival-maximizing attributes (like yield stability and early maturity) and those found in less accessible areas have to be targeted with policies that can enhance the productivity and comparative advantages of traditional varieties.

Farmers' networks and the exchange of information have positively affected not only their perception towards replacement and loss but also the importance they attach to traditional varieties. Thus, to enhance on-farm conservation, mechanisms have to be designed to create and support the exchange of information among farmers so that the farmers who appreciate the utility of traditional varieties (pro-diversity farmers/communities) can share their knowledge and experiences (with other farmers) about traditional varieties.

Farmers who are more experienced with improved varieties and extension services and those who prefer marketable varieties most are the proper partners to be targeted by awareness creation campaigns and mitigation measures.

Farmers' compliance with the demands of on-farm conservation can be enhanced when the conservation measures target those farmers who believe that their livelihoods are more connected to the survival of traditional varieties. On the other hand, localities/farmers/communities of an opposite nature have to be targeted to reverse potential loss.

The results suggest that socio-economic changes (new markets, new crop enterprises, new crop varieties, etc.) and agro-ecological dynamics (climate change, land degradation, drought and desertification) in smallholder farming can play against the comparative advantage of these varieties of crops, especially when traditional varieties retire and become incompatible with the changes. This is a loss that happens in rural areas due to socio-economic and agro-ecological changes occurring in the villages. In those scenarios, measures that enhance the utility of farmers' varieties to the local community can support both farmers' livelihoods and on-farm conservation.

Notes

1   The legal explanations mainly focus on national and international legal/regulatory gaps.
2   Of course, the issue of replacement of traditional varieties by the improved ones is crop-specific. It is more important for crops that have been affected by Green Revolution technologies (e.g. wheat, maize and rice).
3   A review by Brush (1995) for maize, wheat and potatoes in their respective centres of diversity shows that the spread of modern varieties has resulted in changing population structure/compositions, not complete loss.
4   Yifru and Hammer (2006), for instance, have reported a reduction in the use of local varieties over the years for triploid wheat in Ethiopia.
5   Since this is a case study, there was no compelling reason to choose these zones apart from their typical features in terms of facing frequent drought, degraded land and having a mixed cereal-livestock farming system.
6   The response options were: fully agree, partly agree and disagree. For the analysis here, options 1 and 2 are merged into the same category as the reasons given for the choices were the same.
7   This is a process, often called creolization, which has been shown to have happened, for instance, for maize in Mexico (Bellon and Brush, 1994).
8   This refers to the utility of the diversity of traditional varieties of crops (in terms of insurance against yield fluctuations, nutrition and health values of dietary diversity, and other desirable production and consumption attributes) enjoyed by all the farming communities locally.

References

Adesina, A. A. and Baidu-Forson, J. (1995) 'Farmers' perceptions and adoption of new agricultural technology: evidence from analysis in Burkina Faso and Guinea, West Africa', *Agricultural Economics*, no 13, pp1–9

Almekinders, C. J. M., Louwaars, N. P. and de Bruijn, G .H. (1994) 'Local seed systems and their importance for an improved seed supply in developing countries', *Euphytica*, vol 78, no 3, pp207–216

Bardsley, D. K. (2003) 'Risk alleviation via in situ agro-biodiversity conservation: drawing from experiences in Switzerland, Turkey, and Nepal', *Agriculture, Ecosystems and Environment*, vol 99, pp149–157

Bardsley, D. and Thomas, I. (2005) 'In situ agrobiodiversity conservation for regional development in Nepal', *Geojournal*, no 62, pp27–39

Batz, F.-J., Peters, K. J. and Janssen, W. (1999) 'The influence of technology characteristics on the rate and speed of adoption', *Agricultural Economics*, vol 21, no 2, pp121–130

Bellon, M. R. and Brush, S. B. (1994) 'Keepers of maize in Chiapas, Mexico', *Economic Botany*, vol 48, no 2, pp196–209

Bellon, M. R. and Risopoulos, J. (2001) 'Small-scale farmers expand the benefits of improved maize germplasm: a case study from Chiapas, Mexico', *World Development*, vol 29, no 5, pp799–811

Bellon, M. R. and Taylor, J. E. (1993) 'Farmer soil taxonomy and technology adoption', *Economic Development and Cultural Change*, no 41, pp764–786

Brennan, J. and Byerlee, D. (1991) 'The rate of crop varietal replacement on farms: measures and empirical results for wheat', *Plant Varieties and Seeds*, no 4, pp99–106

Brush, S. B. (1991) 'A farmer-based approach to conserving crop germplasm', *Economic Botany*, vol 45, no 2, pp153–165

Brush, S. B. (1995) 'In situ conservation of landraces in centers of crop diversity', *Crop Science*, vol 35, no 2, pp346–354

Brush, S. B., Bellon, M. and Schmidt, E. (1988) 'Agricultural development and maize diversity in Mexico', *Human Ecology*, no 16, pp95–106

Brush, S. B., Taylor, J. E. and Bellon, M. (1992) 'Technology adoption and biological diversity in Andean potato agriculture', *Journal of Development Economics*, no 39, pp65–387

Cleveland, D. A., Soleri, D. and Smith, S. E. (1994) 'Do folk varieties have a role in sustainable agriculture?', *Bioscience*, vol 44, no 11, pp740–751

Cooper, J. C., Lipper, L. M. and Zilberman, D. (2005) 'Synthesis chapter: managing plant genetic diversity and agricultural biotechnology for development', in J. Cooper, L. M. Lipper and D. Zilberman (eds) *Agricultural Biodiversity and Biotechnology in Economic Development*, Springer, New York, pp457–477

Dennis Jr, J. V. (1987) 'Farmer management of rice diversity in Northern Thailand', PhD thesis, Cornell University, Ithaca, NY

Ehrlich, P. R. and Wilson, E. O. (1991) 'Biodiversity studies: science and policy', *Science*, vol 253, no 5021, pp758–762

Hardon, J. J. (1996) 'Conservation and use of agro-biodiversity', *Biodiversity Letters*, vol 3, no 3, pp92–96

Harlan, J. R. (1975) 'Our vanishing genetic resources', *Science*, vol 188, no 4188, pp618–621

Harlan, H. V. and Martini, M. L. (1936) 'Problems and results of barley plant breeding', in US Department of Agriculture, *USDA Yearbook of Agriculture*, US Government Printing Office, Washington, DC, pp303–346

Heisey, P. W. and Brennan, J. P. (1991) 'An analytical model of farmers' demand for replacement seed', *American Journal of Agricultural Economics*, vol 73, no 4, pp1044–1052

Herzon, I. and Mikk, M. (2007) 'Farmers' perceptions of biodiversity and their willingness to enhance it through agri-environment schemes: a comparative study from Estonia and Finland', *Journal for Nature Conservation*, vol 15, no 1, pp10–25

Just, R. E. and Zilberman, D. (1983) 'Stochastic structure, farm size, and technology adoption in developing agriculture', *Oxford Economic Papers*, vol 35, no 2, pp307–328

Matson, P. A., Parton, W. J., Power, A. G. and Swift, M. J. (1997) 'Agricultural intensification and ecosystem properties', *Science*, vol 277, no 5325, pp504–509

Maxted, N., Guarino, L., Myer, L. and Chiwona, E. A. (2002) 'Towards a methodology for on-farm conservation of plant genetic resources', *Genetic Resources and Crop Evolution*, vol 49, no 1, pp31–46

Negatu, W. and Parikh, A. (1999) 'The impact of perception and other factors on the adoption of agricultural technology in the Moret and Jiru Woreda district of Ethiopia', *Agricultural Economics*, vol 21, no 1, pp205–216

Nunes, P., van den Bergh, J. and Nijkamp, P. (2003) *The Ecological Economics of Biodiversity: Methods and Policy Applications*, Edward Elgar, Cheltenham, UK

Perales, H. R., Brush, S. B. and Qualset, C. O. (1998) 'Agronomic and economic competitiveness of maize landraces and in situ conservation in Mexico', in M. Smale (ed.) *Farmers, Gene Banks and Crop Breeding: Economic Analyses of Diversity in Wheat, Maize, and Rice*, CIMMYT and Kluwer Academic Publishers, Boston, MA, pp109–127

Perfecto, I., Rice, R. A., Greenberg, R. and van der Voort, M. E. (1996) 'Shade coffee: A disappearing refuge for biodiversity', *Bioscience*, vol 46, no 8, pp598–608

Perrings, C. and Lovett, J. (1999) 'Policies for biodiversity conservation: the case of Sub-Saharan Africa', *International Affairs*, vol 75, no 2, pp281–305

Perrings, C., Mäler, K.-G., Folke, C., Holling, C. S. and Jansson, B. O. (1995) 'Unanswered questions', in C. Perrings, K.-G. Mäler, C. Folke, C. S. Holling, and B. O. Jansson (eds) *Biodiversity Loss: Economic and Ecological Issues*, Cambridge University Press, Cambridge, UK, pp301–308

Potvin, C., Owen, C. T., Melzi, S. and Beaucage, P. (2005) 'Biodiversity and modernization in four coffee-producing villages of Mexico', *Ecology and Science*, vol 10, no 1, pp18

Powledge, F. (1998) 'Biodiversity at the crossroads', *BioScience*, vol 48, no 5, pp347–352

Smale, M. (1997) 'The green revolution and wheat genetic diversity: some unfounded assumptions', *World Development*, vol 25, no 8, pp1257–1269

Smale, M. and Heisey, P. W. (1995) 'Maize of the ancestors and modern varieties: the microeconomics of high-yielding variety adoption in Malawi', *Economic Development and Cultural Change*, vol 43, no 2, pp351–368

Smale, M., Just, R. E. and Leathers, H. D. (1994) 'Land allocation in HYV adoption models: an investigation of alternative explanations', *American Journal of Agricultural Economics*, vol 76, no 3, pp535–546

Srivastava, J., Smith, N. J. H. and Forno, D. (1996) *Biodiversity and Agriculture: Implications for Conservation and Development*, World Bank Technical Paper no 321, World Bank, Washington, DC

Subedi, A., Chaudhary, P., Baniya, B. K., Rana, R. B., Tiwari, R. K., Rijal, D. K., Sthapit, B. R. and Jarvis, D. I. (2003) 'Who maintains crop genetic diversity and how? Implications for on-farm conservation and utilization', *Culture and Agriculture*, vol 25, no 2, pp 41–50

Thrupp, L. A. (2000) 'Linking agricultural biodiversity and food security: the valuable role of agro-biodiversity for sustainable agriculture', *International Affairs*, vol 76, no 2, pp265–281

Tilman, D. (1998) 'The greening of the green revolution', *Nature*, vol 396, pp211–212

Tripp, R. (1996). 'Biodiversity and modern crop varieties: sharpening the debate', *Agriculture and Human Values*, vol 13, no 4, pp48–63

Virchow, D. (ed.) (2003) *Efficient Conservation of Crop Genetic Diversity: Theoretical Approaches and Empirical Studies*, Springer-Verlag, Berlin

Wale, E. (2004) 'The economics of on-farm conservation of crop diversity in Ethiopia: Incentives, attribute preferences, and opportunity costs of maintaining local varieties of crops', PhD thesis, University of Bonn, Bonn, Germany

Wale, E. (2006) 'What do farmers financially lose if they fail to use improved seeds? Some econometric results for wheat and implications for agricultural extension policy in Ethiopia', *Ethiopian Journal of Economics*, vol 12, no 2, pp59–79

Wale, E. (2008) 'Challenges in genetic resources policy making: Some lessons from partic-ipatory policy research with a special reference to Ethiopia', *Biodiversity and Conservation*, vol 17, no 1, pp21–33

Wale, E., Chishakwe, N. and Lewis-Lettington, R. (2009) 'Cultivating participatory policy processes for genetic resources policy: lessons from the Genetic Resources Policy Initiative (GRPI) project', *Biodiversity and Conservation*, vol 18, no 1, pp 1–18

Wolff, F. (2004) 'Legal factors driving agrobiodiversity loss', *Environmental Law Network International Review*, vol 1, pp1–11

Wood, A., Stedman-Edwards, P. and Mang, J. (2000) *The Root Causes of Biodiversity Loss*, Earthscan, London

Wood, D. and Lenne, J. M. (1997) 'The conservation of agro-biodiversity on-farm: questioning the emerging paradigm', *Biodiversity and Conservation*, vol 6, no 1, pp109–129

Yifru, T. and Hammer, K. (2006) 'Farmers' perception and genetic erosion of tetraploid wheat landraces in Ethiopia', *Generic Resources and Crop Evolution*, no 53, pp1099–1113

Zander, K. K. and Drucker, A. G. (2008) 'Conserving what's important: using choice model scenarios to value local cattle breeds in East Africa', *Ecological Economics*, no 68, pp34–45

Zander, K. K., Drucker, A. G. and Holm-Müller, K. (2009) 'Costing the conservation of animal genetic resources: the case of Borana cattle in Ethiopia and Kenya', *Journal of Arid Environments*, vol 73, nos 4–5, pp550–556

Part 3

Market Value Chains, Commercialization and On-farm Conservation Policy

Chapter 5

Consumers' Attribute Preferences and Traders' Challenges Affecting the Use of Local Maize and Groundnut Varieties in Lusaka: Implications for Crop Diversity Policy

E. Kuntashula, Edilegnaw Wale, J. C. N. Lungu and M. T. Daura

Summary

There has been inadequate information that presents the impact of consumers' and traders' preferences on the production and utilization of local crop varieties. Recognizing the links among consumption, production, trade and on-farm utilization, this chapter deals with the preferences of urban consumers and traders by taking maize (Gankata) and groundnut (Kadononga) as illustrative examples. Obviously, consumers and traders' preferences are important to the extent that farmers produce maize and groundnut not only for themselves but also for the market. The overall objective was to generate information that can support better conservation outcomes through increased production and utilization of tradi-tional crop varieties. This is done with due consideration for those varieties with no current consumption or trading utility but of potential future insurance value to sustainable agriculture.

The study took place in urban Lusaka and used structured questionnaires to capture information from 106 and 63 groundnut consumers and traders, respec-tively, and 104 and 60 maize grain consumers and traders, respectively. The data are analysed using variety attribute ranking and regression analysis.

The results have revealed that before making decisions to buy maize, consumers pay more attention to quantity-related attributes (such as grain size and kernel density) than to quality-related attributes (such as food taste). In contrast, when consumers make decisions to buy groundnut, both quantity and quality attributes were perceived as important. This is because, unlike groundnut which is a non-staple food crop, maize is a staple food in Zambia, taking the lion's share of the income of urban poor consumers.

The results imply that, when dealing with staple foods such as maize, breeders should focus on quantitative traits whereas both quantitative and qualitative traits should be considered for non-staple food crops. Since the relevance to consumers of local maize and groundnut varieties is linked to presence or absence of the preferred traits, genetic resources conservation policy should integrate varieties with other attributes which may not be currently valuable to consumers but of potential public utility in the future. The conservation strategies have to be linked to farmers' and trader's attribute preferences. That is how policy can achieve the objectives of both agro-biodiversity conservation and enhanced livelihoods.

Introduction

The local crop varieties constitute a valuable component of the available genetic diversity necessary in the development and improvement of crop enterprises. However, this diversity is continuously being lost or threatened because of natural- and human-driven factors. Local crop varieties are often underutilized and there is often little effort made in terms of the realization of their contribution to the socio-economic needs of the people locally and nationally. In addition to their contribution to risk reduction in agriculture, local varieties are also known to have potential nutritional values and cooking qualities.

Maize is Zambia's staple food and Gankata is a widely grown local maize variety among most small-scale farmers in Zambia. Alongside maize, groundnut features predominantly in the production systems of Zambian small-scale farmers. Kadononga, a local traditional groundnut, is a very common variety in the country. During a priority-setting exercise conducted under the auspices of Genetic Resource Policy Initiative (GRPI) project in 2006, Gankata and Kadononga were found to be the most widely grown local varieties in Zambia. However, the area allocated to these crop varieties was found to be significantly smaller than that for hybrid varieties of maize and groundnut (GRPI, 2006). Generally, the production of local crop varieties (including maize and ground-nut) has been declining over the years. According to Gumbo (1986), the production of local maize varieties has been decreasing since Zambia's independence in 1964. This is mainly attributed to the emphasis the then govern-ment had attached to improving the existing genotypes of maize such as 'improved SR52', which was introduced in 1979. During the time, SR52 was developed to meet a wide range of farmers' needs in different environmental conditions (Gumbo, 1986).

Over time, the use of improved maize varieties in Zambia has been increasing. Howard and Chitalu (2000), for instance, note that the area allocated to improved seeds increased in Zambia from 4.5 per cent in 1983 to 58.6 per cent in 1992. The use of local maize varieties reduced from 65.5 per cent to 26.0 per cent during the same period. More notably, the introduction of a more liberalized market economy (in 1991) brought with it many seed companies whose major objective was profit maximization through the promotion of their improved crop varieties. In addition, there has been a lack of appreciation or inadequate integration of local resources issues (including crop genetic resources) into various sectoral policies and instruments. All these factors have led to drastic drops in the production of local crop varieties in the country.

The displacement of local crop varieties by improved ones in rural communities has been the result of the advantages of improved varieties in production attributes (such as high yields and early maturity). Several studies (for instance, Mafuru et al, 1999) have shown the main production advantages that improved varieties have over local varieties. In this case study chapter, it is recognized that both production attributes (e.g. yield), market or trade attributes (e.g. price) and consumption attributes (e.g. taste) play an important role in crop variety use decisions. While there is increasingly important literature on variety-specific attributes with impact on the production of local varieties (for instance, Smale et al, 2001; Edmeades et al, 2004; Wale and Yalew, 2007), there is inadequate or little information on the variety attribute preferences of consumers and traders. Most previous studies are on farmers' preferences which naturally focus largely on production traits (Wale et al, 2005). Since many farmers produce for their families and the market, consumers' and traders' variety attribute preferences affect the production of local crop varieties on-farm. Variety choice is a revealed preference to respond to production constraints, satisfy consumption preferences and fulfil specific market requirements (Smale et al, 2001). Since farmers' production decisions and consumers' preferences are linked through the market, farmers respond not only to their own preferences but also to consumers and traders' demands/preferences. This consumption/production linkage can be explored (and opportunities seized) to inform policy decisions.

Understanding important attributes affecting consumers and traders' preferences for local varieties can shed light on the missing links in the market chain and marketing constraints that affect the consumption and sustainable use of traditional varieties of crops. In most of the existing literature, there is a systematic omission on the role consumers and traders play in the maintenance and/or disappearance of certain local varieties.

In this case study, it is assumed that production of local varieties is affected by the preferences of urban consumers and the market constraints the traders face. As the empirical results will show later, this assumption has been shown to hold in the areas studied. The study uses maize (Gankata) and groundnut (Kadononga) to understand the attributes that consumers like, or otherwise, and the marketing constraints that traders face in the marketing of the two products. The overall objective is to study consumers and traders' variety preferences and generate information that can support decision-making towards the sustainable use and conservation of crop genetic resources. The study will also explore the mechanisms for identifying traditional varieties that have less chance to stay on-farm – to support possible conservation priority-setting.

The next section presents the methodology, describing the data collection procedures and methods of data analysis. This is followed by the presentation of findings from both maize and groundnut trader and consumer surveys. The major results are then discussed. Finally, the conclusions and implications of the results are presented, drawing from the empirical results.

Methodology

Study areas

The study was conducted in Lusaka, the most densely populated city in Zambia (CSO, 2000). There are markets with large volumes that involve diversity of various products including most agricultural products. Interviews among groundnut (Kadononga) consumers took place in seven residential areas, namely Mtendere, Ng'ombe, Kalingalinga, Kalikiliki, Chawama, Mandevu and Chipata. There was no stratification made in sampling residential areas and groundnut consumers. For maize, consumers were not stratified according to residential areas because only a small proportion of Lusaka's urban consumers directly buy maize grain for subsequent milling as straight-run meal (Kuntashula, 1999). Most consumers buy maize flour. Consumers of maize grain were interviewed either at the place of grain purchase (urban markets) or at maize grain processing locations (at hammer millers). A total of the household heads of 106 groundnut consumers and 63 groundnut traders were interviewed for the groundnut survey. A total of 104 consumers and 60 traders were interviewed for the maize grain survey. None of the interviewed consumers or traders was a maize or groundnut producer. It has to be re-emphasized that this is a case study and no claim can be made on the representativeness of the sample. The results only hold in those contexts where the features of the study areas and the interviewed consumers/traders prevail.

Data collection and description

Structured questionnaires were used throughout the study. During pre-testing of the questionnaires on 20 households, 5 main maize attributes (market price, grain colour, grain size, kernel density and food taste) were identified as relevant for consumers. Similarly, a total of 8 important attributes were identified for groundnut (market price, grain colour, easiness to peel testa, kernel density, food taste, oil content, grain size and number of pods).

All the interviewed maize consumers were asked to rate all the attributes on a scale of 1 to 5 regardless of whether or not they viewed them as important in their decision to buy maize grain because all attributes were important for all maize consumers. During the groundnut interviews, consumers were asked to only rate the attributes they found relevant in their groundnut-buying decisions because some attributes were not important for some groundnut consumers. In both

Table 5.1 *Data description, continuous explanatory variables*

Variable	Variable explanation	Mean (SD)
HHsize	Household size	5.10 (1.58)
Agehh	Age of the household head (years)	36.90 (7.70)
EducHH	Education level of household head (grade)	1.55 (0.65)
HHinc	Total household monthly income (Zambian Kwacha[1])	474,327 (243,735)
NoRooms	Number of rooms in the house	2.36 (0.89)
Freqpur	Frequency of purchase per month	8.07 (4.60)
Nearmrt	Time taken to walk to the nearest maize market (minutes)	24.96 (18)

Source: GRPI Zambia, consumer data, 2007

cases, a rate of 1 meant least important and a rate of 5 meant an attribute to which consumers attach the highest importance when making purchasing decisions. In between were rates of 2, 3 and 4, designating the respective importance of the attributes in an orderly manner.

The questionnaires had sections covering individual characteristics, ground-nut and maize grain buying and selling activities, marketing constraints, asset ownership and the market-related decision-making process. The description of the data (from maize consumers) used in the regression analysis is presented in Tables 5.1 and 5.2.

Table 5.2 *Data description, dummy explanatory variables*

Variable	Description	Label	%
Gender	Gender of the household head	Male (1)	70.2
		Female (0)	29.8
Married	Marital status of household head	Married (1)	72.1
		Not married (0)	27.9
Whenbuy	Time of the year Gankata is bought	Right after harvest (1)	72.1
		Other time or throughout the year (0)	27.9
HHowners	Household house ownership	Yes (1)	66.3
		No (0)	33.7
Purp_maz	Purpose for which maize is bought	Mealie meal (1)	79.8
		Mealie meal and 'samp'[2] (0)	20.2
Evenspr	Is purchase of maize evenly spread?	Yes (1)	76.9
		No (0)	23.1
Buylm	Do you have a chance to buy local maize whenever you want?	Yes (1)	72.1
		No (0)	27.9
PreferLocal	Do you prefer the local maize to hybrid maize?	Yes (1)	87.5
		No (0)	12.5

Source: GRPI Zambia, consumer data, 2007

According to the data in the preceding tables, most household heads were male and married. The household size was about 5 people per household. The average consumer monthly income is about 474,327 Zambian Kwacha (≈US$118.6). The average number of rooms in the house is about 2.4 while the frequency of purchase of maize per month is about 8. It takes about 25 minutes to walk to the nearest market for a typical consumer.

Data analysis

The survey data were descriptively analysed using SPSS (version 11) and the regression analysis was done using STATA (version 9). The attribute preference levels were subjected to paired t-test analyses while regression analysis was conducted on the influence of different household contextual characteristics on the quantity of maize purchased per week.

Describing consumers' purchasing decisions and preferences for maize

The majority of the interviewed households (86 per cent) do not consume Gankata evenly due to scarcity during some months of the year. Most consumers prefer Gankata for large grain size (47 per cent), kernel density (28 per cent) and good taste (19 per cent). That is why they continue to buy and use Gankata in the presence of other varieties such as Pioneer and Moffat. However, the quantity of Gankata coming to the market is declining due to the decline in production.

Comparing the last five years, the purchase of local maize varieties has decreased for most consumers (60 per cent) due to its scarcity and expensive price, which has forced most consumers to reduce their purchase. For the households who had increased purchases of local maize grain, increase in family size was cited as the major reason. Trader enquiries, however, revealed that price differences between local and improved maize was hardly significant and in most cases nonexistent. This is because the government regulates the maize market at the source and sets the same price regardless of the quality differences among the different types of maize grain. During the survey, the price of 50kg of maize was around K40,000 (≈US$10) in almost all the markets. Essentially, the same maize market price prevailed in most markets surveyed, regardless of the variety of maize.

The purchase of improved maize in the last five years has increased for 50 per cent of the consumers due to ready availability while it has decreased for others due to poor taste. Most consumers (81 per cent) buy maize from vendors. In terms of time of purchase, most consumers (80 per cent) buy maize after harvest (May–November).

According to the consumers, good quality maize is one which: is white in colour, has large grain size and has good kernel density. The most important characteristics considered by consumers in the purchase of maize are grain colour, grain size, kernel density, market price and food taste (in the order presented). In this connection, the most important desirable consumption traits of

Table 5.3 *Consumers' preferences of the local to improved maize grain (% households)*

	Choice	Grain size	Colour	Taste	Kernel density	Taste and kernel density	Readily availability
				Reason			
Do you prefer	Yes	27.9	5.8	9.6	31.7	12.5	0
local to	No	5.8	1.0	1.0	1.0	0	3.8
improved maize?							

Source: GRPI Zambia, consumer data, 2007

Gankata are: quantity of flour, better taste and white flour. A few households (around 12 per cent) have pointed out that Gankata had undesirable attributes, such as its being expensive, having poor taste and/or an unpleasant colour. Table 5.3 reports consumers' preferences between local and improved maize grain with respect to different variety traits.

About 88 per cent of the households preferred local to improved maize grain with respect to all the variety traits (Table 5.3). The superiority of the local varieties in terms of kernel density, grain size, food taste and colour were frequently cited as the major reasons for this preference.

Attributes relevant for maize grain purchase decisions

Maize in Zambia is consumed in the form of a thick porridge locally known as 'nshima'. Almost all consumers (99 per cent) purchasing dry maize grain mill it into flour to make nshima. In addition, about 20 per cent of consumers indicated that they purchase maize grain for use as 'samp'. Table 5.4 reports the relevance of different maize variety traits for the interviewed consumers.

Grain size stood out as the most important attribute for a household's maize grain purchase decisions. More than 80 per cent of consumers attached a 'relevant' or 'very relevant' ordinal score to grain size. Most households considered food taste and market price as less relevant traits (Table 5.4). Statistical tests on the means of the ratings showed that the ratings of the attributes in order of importance were: grain size (4.23) > kernel density (3.46) = grain colour (3.15) > market price (2.09) = food taste (2.01). The mean paired rank differences for these attributes are shown in Table 5.5.

Grain colour, grain size and kernel density were rated higher than the grain market price. Grain size was rated higher than grain colour, kernel density and food taste. Kernel density and grain colour were rated higher than food taste (Table 5.5). All these results confirm that for staple food crops like maize, quantity traits stand out as most important for consumers' purchase decisions.

Table 5.4 *Rating attributes relevant for consumers in maize purchase decisions (N = 104)*

Attribute	Very low relevance	Low level of relevance	Medium level of relevance	Relevant	Very relevant
Market price	42.3	23.1	20.2	12.5	1.9
Grain colour	4.8	29.8	26.0	24.0	15.4
Grain size	2.9	3.8	10.6	32.7	50.0
Kernel density	6.7	16.3	26.0	26.0	25.0
Food taste	46.2	26.0	14.4	4.8	8.7

Source: GRPI Zambia, consumer data, 2007

Eighty-eight per cent of the Gankata consumers agreed that Gankata possesses the most desirable maize variety attributes and that was the reason why consumers accepted it in the market. That also explains why this variety has stayed in the market for so long and it is well known by producers, consumers and traders.

About 95 per cent agreed that the quantity of Gankata coming to the market has been declining in recent times. About 72 per cent of the consumers have indicated that they could not always find local maize whenever they wanted to buy it. However, 87 per cent of the consumers have pointed out that they could always find improved maize varieties whenever they wanted to buy them. This is despite consumers' preference (88 per cent) for local maize.

A few consumers (17.3 per cent) had memories of local maize varieties that are fast disappearing from the market. The varieties they mentioned include Kalimwa, Kapwawangu, Nsanga, Kanjerenjere, Bantam-Chipata, Senga and Manidza Chala.

Table 5.5 *Pair-wise comparisons of attributes considered important for purchase of maize grain*

	Paired differences		t	Degrees of freedom	Significance (2 tailed)
	Mean	Standard error			
Market price – Grain colour	−1.07	0.18	−6.04	103	0.00
Market price – Grain size	−2.14	0.16	−13.24	103	0.00
Market price – Kernel density	−1.38	0.18	−7.48	103	0.00
Grain colour – Grain size	−1.08	0.14	−7.53	103	0.00
Grain colour – Food taste	1.15	0.19	5.94	102	0.00
Grain size – Kernel density	0.77	0.17	4.45	103	0.00
Grain size – Food taste	2.21	0.19	11.94	102	0.00
Kernel density – Food taste	1.48	0.17	8.65	102	0.00

Note: Only significant differences are reported.
Source: GRPI Zambia, consumer data, 2007

Attributes relevant for groundnut purchase decisions

As in the case of maize, consumers were asked to order their preferences for different attributes of groundnut varieties. Table 5.6 reports the results of this exercise. Most groundnut consumers consider food taste and grain size as most relevant traits in their decisions to purchase groundnut.

Kernel density, grain colour and oil content were cited as important attributes next to food taste and grain size. Number of pods was the most irrelevant attribute considered by groundnut consumers because most consumers buy shelled grains of groundnut. The other most irrelevant attributes were easiness to peel testa and groundnut market price (Table 5.6).

The ratings above suggest that consumers' preferences are in the order of: food taste = grain size > oil content = kernel density = grain colour > market price = easiness to peel > number of pods. The mean paired rank differences for groundnut attributes are reported in Table 5.7.

The pair-wise ratings showed that, with the exception of easiness to peel (and number of pods), all other attributes were viewed more important than the market price attribute. While grain colour was found to be more important than easiness to peel, food taste and grain size were rated more important than grain colour. Food taste, oil content and grain size were all viewed as more important than easiness to peel. Food taste and grain size were rated more important than kernel density. Grain size ratings dominated oil content ratings, which, in turn, dominated food taste ratings (Table 5.7).

Although 28.8 per cent of consumers did not agree, 58.5 per cent agreed that Kadononga possesses most of the desirable attributes of groundnut. About 74 per cent agreed that the quantity of Kadononga coming to the market these days has been declining. Most consumers (83 per cent) had an opportunity to buy local groundnut whenever they wanted. Regarding preferences of consumers for

Table 5.6 *Rating the releveance of attributes for consumers in groundnut purchase decisions (N = 106)*

Attribute	Very low relevance	Low level of relevance	Medium level of relevance	Relevant	Very relevant	Not relevant at all
Market price	9.4	17.0	7.5	2.8	1.9	61.3
Grain colour	9.4	9.4	17.0	11.3	14.2	38.7
Easiness to peel testa	13.2	3.8	1.9	0.9	0	80.2
Kernel density	6.6	17.0	18.9	13.2	7.5	36.8
Food taste	0.9	8.5	19.8	24.5	40.6	5.7
Oil content	3.8	16.0	15.1	15.1	7.5	42.5
Grain size	4.7	8.7	13.2	31.1	33.0	8.1
Number of pods	1.9	0	0	0	0	98.1

Source: GRPI Zambia, consumer data, 2007

Table 5.7 *Pair-wise comparisons of attributes considered important for purchase of groundnut*

	Paired differences		t	Degrees of	Significance
	Mean	Standard error		freedom	(2 tailed)
Market price – Grain colour	–1.17	0.31	–3.75	29	0.001
Market price – Kernel density	–2.00	0.46	–4.34	11	0.001
Market price – Food taste	–1.95	0.21	–9.24	38	0.000
Market price – Oil content	–0.81	0.32	–2.55	15	0.022
Market price – Grain size	–1.77	0.26	–6.83	34	0.000
Grain colour – Easiness to peel	2.25	0.62	3.63	7	0.008
Grain colour – Food taste	–0.73	0.25	–2.93	59	0.005
Grain colour – Grain size	–0.65	0.23	–2.82	59	0.006
Easiness to peel – Kernel density	–1.69	0.38	–4.43	12	0.001
Easiness to peel – Food taste	–2.50	0.26	–9.75	19	0.000
Easiness to peel – Oil content	–1.71	0.36	–4.72	16	0.000
Easiness to peel – Grain size	–2.05	0.39	–5.30	19	0.000
Kernel density – Food taste	–0.94	0.24	–3.88	62	0.000
Kernel density – Grain size	–0.97	0.22	–4.34	64	0.000
Food taste – Oil content	0.84	0.23	3.65	57	0.001
Oil content – Grain size	–0.72	0.26	–2.81	53	0.007

Note: Only significant differences are reported.
Source: GRPI Zambia, consumer data, 2007

groundnut varieties, about 75 per cent of the respondents preferred local varieties to improved ones. About 21 per cent of consumers could remember some local groundnut varieties that had disappeared from the market, including Kayoba and Sorontone.

Factors affecting the quantity of maize grain purchased

In addition to maize variety attributes, consumption could be influenced by consumers' household characteristics. To explain the quantity of maize purchased by urban consumers, OLS (ordinary least squares) regression is conducted using the variables explained in Tables 5.1 and 5.2. The response variable is the average quantity of maize bought (kg) per week. The descriptive statistics show that, on average, a household purchased about 44kg per week.

The following regression results are heteroscedasticity corrected. The discussions on the expected relationships and the regression results are presented below for the significant variables.

Gender of the household head

It was expected that male-headed households could buy more quantities of Gankata than female-headed households. The results show that the male-headed households bought 6.1kg more Gankata per week than their female-headed counterparts, other factors held constant. This could be attributed to the fact that male-headed households are economically better-off.

Household size

It can be hypothesized that the more members a household has the greater quantity of Gankata they are likely to buy because maize is generally a staple food in Zambia. Every member of household would want to consume maize meal at least once in a day. Since Gankata is in most cases bought for maize meal, it is not surprising that household size increases the quantity of Gankata that the house-hold is likely to buy. The regression results suggest that an additional household member will trigger a household to buy 4.2kg more of Gankata per week.

Table 5.8 *Regression results on factors affecting average quantity of Gankata bought, 2007*

Variable	Coefficient (t)
Gender	6.11^* (1.75)
HHsize	4.26^{***} (6.82)
Agehh	0.27^* (1.61)
EducHH	1.02 (0.60)
Married	-7.40^{**} (−2.07)
HHinc	−4.34e-07 (−0.08)
NoRooms	2.45^* (1.75)
Whenbuy	−0.68 (−0.28)
HHowners	-8.28^{***} (−3.65)
Purp_maz	-5.08^* (−1.80)
Freqpur	−0.15 (−0.57)
Evenspr	-4.67^* (−1.59)
Nearmrt	0.10^* (1.67)
Buylm	-4.50^* (−1.64)
PreferLocal	-4.99^* (−1.70)
Constant	25.94 (2.21)
Number of obs = 104	$F_{(15, 88)} = 9.42$
Prob > F = 0.000	$R^2 = 0.49$
Adj R^2 = 0.401	

Notes: *** significant at 1% level; ** significant at 5% level; and * significant at 10% level. Values in parentheses are the ratio of the coefficient to the estimated asymptotic standard error.
Source: GRPI Zambia, consumer data, 2007

Marital status of the household head

Households with married heads bought more Gankata than single-headed households. According to the results, the single-headed households bought 7.4kg less Gankata per week than their married counterparts. This can be attributed to the relative income poverty of single-headed households.

Number of rooms and house ownership

The regression results suggest that households with more rooms in their houses were buying more Gankata than those with fewer rooms. The number of rooms in a house is often positively related to the level of income and living standard. Similarly, those who own houses are also relatively better-off than those who don't own houses. Consequently, the results show that house ownership positively affects the quantity of Gankata purchased.

Purpose of purchasing maize

Maize grain is mainly bought for making either maize flour that is later used to make nshima, the staple food for most Zambians, or maize samp, mainly used as a breakfast meal. The quantities of Gankata bought were significantly related to the purpose for which maize grain was bought, i.e. households whose main purpose of buying maize was nshima were buying more Gankata. Maize grain is mostly bought by the poor to produce nshima (Kuntashula, 1999).

Time taken to reach the nearest maize market

According to the regression results, the shorter the time to reach the nearest maize market, the more quantities of Gankata were purchased, other factors held constant. For households closer to markets, the transaction cost is lower and they do not have to exert too much effort to obtain Gankata. This result supports previous studies whereby distance to markets for local maize varieties is found to be a serious constraining factor in Zambia as most local maize varieties are consumed in places where they are produced (Hara, 2008).

Preference to local maize

The results suggest that those consumers who prefer local to improved maize varieties bought more Gankata. This is expected as Gankata is one of the most common and well-known local maize varieties.

Characteristics of maize grain and groundnut traders

The interviewed maize grain traders had their grain trade business located in several areas of urban Lusaka including Soweto, Mandevu, Kalingalinga, Mtendere, Chipata, Kalikiliki, Chawama, George, Ufulu, Ng'ombe, Katambala,

Kadzimai, Chaisa and Lupilu. Groundnut traders have their businesses in Soweto, Buseko, Kuku, Mtendere and Chawama.

About half of the sampled maize grain traders were male and the average age of the traders was 37 years. Most of maize grain traders (65 per cent) had completed secondary schooling. While most traders were trading maize alone, only a few were trading other cereals such as popcorn (21.8 per cent) and sorghum (1.8 per cent). About 33.3 per cent of the maize traders were involved in non-cereal trading.

About two-thirds of the groundnut traders are male. The average age of the groundnut traders was 33.4 years. About 50 per cent, 45 per cent and 5 per cent of the groundnut traders had completed primary schooling, secondary schooling and tertiary education, respectively. While 90 per cent of the groundnut traders were middlemen, the rest were farmers. About 57 per cent of the groundnut traders run other businesses in addition to groundnut business. The other types of businesses that these traders are engaged in include grocery, beans, poultry, vegetables, beans, kapenta (dried sardines) and cloths.

Maize grain trading and traders' preferences

Maize trading in most urban Lusaka markets is dominated by the seasonal petty traders. The traders buy maize grain in the rural areas of the country and bring it to urban areas for resell. Usually these traders are involved in other businesses after the maize grain is out of season. There is a high supply of maize just after harvest, i.e. starting from the end of May through June to September. The supply begins to dwindle towards the end of the year as the rainy season approaches. In January to April there is a serious maize grain scarcity.

Most maize traders (more than 90 per cent) revealed that maize grain supply is high just after harvest and low during the rainy season. This also shows that traders do not target to supply maize when maize supply from the farmers is low. About 48 per cent of the maize traders indicated that it only took up to three days to dispose of the maize grain while the rest were able to do so within a week.

All the maize traders interviewed sell Gankata. The majority of the traders (78.3 per cent) indicated that the selling of Gankata was seasonal, which was mainly attributed to lack of availability throughout the year.

Most of the traders (61 per cent) revealed that local maize was preferred to hybrid maize and a few traders (17 per cent traders) mentioned that Gankata was the most preferred maize variety. Most traders attributed the preference to local maize in general and Gankata in particular to the varieties being demanded by most consumers. They revealed that consumers like local varieties because the varieties produce a lot of mealie meal (more flour) and can be used to make good samp.

The major challenges faced by the maize grain traders in order of their importance were: shortage of capital, low grain prices, high transportation costs, lack of supply, lack of demand and poor market infrastructure (e.g. storage costs and lack of information).

Groundnut trading and traders' major challenges

As it is the case for maize, groundnut traders revealed that supply of groundnut (including Kadononga) is high just after harvest and Kadononga is mainly traded during harvest season and hardly traded throughout the year. The supply is normally low during the growing season and seasonality is the common feature of trading groundnut.

The major challenges faced by the groundnut traders were high transportation costs, seasonality in supply, lack of capital and market-related problems (low demand, poor market infrastructure, price fluctuation and high market levies). The problem of seasonality in supply was further compounded by the absence of storage facilities and high storage costs.

Synthesis and discussions

Maize and groundnut feature prominently in the consumers' food basket in Zambia. Maize is Zambia's staple food while groundnut is commonly eaten as a roasted snack, added as a relish to maize meal porridge. As a staple food, more than half of Zambia's population relies on maize for most of their calorie intake and spend a large share of their disposable income on it (Kuntashula, 1999).

Given the importance attached to maize, it is expected that the quantity attributes that are directly related to its availability are likely to dominate the other traits. In this study, consumers rated grain size high (followed by kernel density), showing that maize consumers first attach importance to issues of quantity before looking at the quality attributes like food taste. Regression results also show that the households' characteristics such as family size (with direct implication on market demand) influenced the quantities of maize bought. This again shows that availability of maize is far more important than quality, especially for many poor urban consumers.

Food taste and market price were rated low by maize grain consumers. Although there are some variations in the way the meals from different maize grain would taste, these differences have not been important to most of the urban poor who make the maize meals using hammer or custom mills. One of the reasons for the urban poor to use custom mills is to obtain more meal than the quantity of maize flour the same amount of money would buy in the supermarkets and shops (Hassan, 2005: Kuntashula, 1999).

The fact that market price is irrelevant in the maize grain purchase may look counter-intuitive to economic theory. It must be noted, however, that the floor price for maize grain in the country is often set by the government to protect the producers. The urban market price is generally the same in urban markets irrespective of the quality differences.[3] The floor price of maize will be the same whether the variety is local or improved, yellow or white or multicoloured. This explains why consumers did not attach much value to market price as an attribute in their decision to purchase one particular maize grain type to another. Of course, price as a factor becomes important in deciding whether to purchase grain to be milled later or to purchase meal that has already been industrially milled. Most studies (e.g. Rubey and Masters, 1994; Bagachwa, 1994) have indicated that the relatively low price of hammer (custom) milled meal has forced most of the urban poor to resort to hammer mills. Since the interviewed maize grain consumers are from relatively poor areas of Lusaka, the price of maize and/or maize meal will be an important consideration in the choice of meal type, not maize type.

The most favoured attribute for Gankata is grain size. Most consumers agreed that Gankata had most attributes desirable for maize and that was the reason why consumers accepted it in the market. In addition, Gankata has a good taste for food (GRPI, 2006). Although most consumers preferred local varieties (like Gankata) to improved varieties, they alluded to the fact that the number of other local varieties has been declining over the years. Most of these varieties are small-seeded and have less flour compared to Gankata, which could be the reasons why they have disappeared from the market.

Unlike maize, most groundnut consumers give more priority to issues of quality (such as food taste and oil content) in making decisions to buy groundnut. Size of groundnut grains and the prevailing market prices were not important for consumers. Since it is not a staple food, consumers could afford to prefer food taste to quantities of groundnut. According to most consumers, Kadononga possesses most of the desirable attributes of groundnut.

Maize grain and groundnut, including local varieties of these crops, are widely traded on several Lusaka urban markets. Generally, there was no evidence suggesting that hybrid varieties are more favoured than local varieties. In fact, there was some evidence pointing to the fact that local maize varieties are preferred by traders because the varieties are also preferred by most customers. This case study has shown that the marketing of both maize grain and groundnut is highly seasonal. The two products are in high supply just after harvest and then they will become rare commodities during the rain or growing season. Most traders of these products are not involved in storage (due to capital shortage) and hence a continuous supply of the commodities beyond a few months after harvest is not occurring. In addition, high transportation cost has reduced the volumes of local maize grain and groundnut marketed in the urban market.

Conclusions

The results of this chapter have shown that grain size followed by kernel density were rated high by maize consumers, and food taste and market price were rated low. Thus, for staple food crops like maize, consumers attach importance to issues of quantity before they start to value quality attributes like food taste. The direct and strong impact of household characteristics such as family size on quantities of maize bought further demonstrates the importance of quantity and availability to consumers. The most favoured attributes of Gankata are grain size and kernel density. Most consumers agreed that Gankata has these desirable attributes. That is the reason why this well-known variety has prevailed in the market for so long as it satisfies quantity characteristics that consumers demand most.

The chapter also demonstrates some important contrasts between maize and groundnut. For groundnut grain, the traits which received more preference by consumers were food taste, oil content and grain colour (quality attributes). Market price, easiness to peel and the number of pods are the least important for groundnut. Unlike maize, quality traits are more important for groundnut, a non-staple food crop.

According to most consumers, Kadononga, the most common traditional groundnut variety, possesses most of the desirable attributes of groundnut, which again explains its survival for so long in the market and on farmers' fields. The availability of the most desired attributes embedded in the local crop varieties (such as Gankata and Kadononga) will ensure that these varieties continue to be demanded by consumers, marketed by traders and grown by farmers. Consumers, however, have observed the general trend that the supply of Kadononga and other local crop varieties have been declining in the urban markets. This trend of potential policy concern might require further investigation in the future, especially for sustainable use and conservation of traditional crop varieties.

Local varieties of maize grain and groundnut are widely traded on several Lusaka urban markets. In fact, there was some evidence pointing to the fact that local maize varieties are preferred by traders due to greater demand by many consumers. The study has shown that the marketing of both maize grain and groundnut is highly seasonal, i.e. the two products are in high supply just after harvest and become rare commodities during the rain or growing seasons.

The results of this chapter show that the income/poverty status of consumers, proximity to markets, purpose of purchasing maize for the staple food (nshima prepared from mealie meal or maize grain powder) and consumers' preference for local varieties over improved ones are the most important factors to explain the quantity of maize (Gankata) purchased.

Implications for policy

What do the results mean for the conservation and sustainable utilization of traditional varieties of the two crops studied? There are some important policy implications of the results for breeding, production and utilization of traditional varieties of maize and groundnut. However, as this chapter is a case study, the results cannot be extrapolated beyond similar study areas.

The results suggest that future breeding efforts should target quantitative traits (like kernel density and grain size) for staple food crops like maize and both quantitative and qualitative traits for non-staple food crops like groundnut. Policy-makers should take advantage of the already existing better preference of local varieties (over hybrid or improved ones) by consumers to further increase their demand by supporting production, processing and marketing of traditional varieties of crops. Campaigns and seed fairs that take advantage of the good attributes of local varieties should be mounted. Deliberate efforts to commercialize local varieties can ensure the continuous survival of these varieties on farmers' fields and improve conservation outcomes. These measures could increase urban demand for local maize and groundnut and thereby increase their production, use and on-farm conservation. For better conservation and livelihoods outcomes, traditional varieties of these crops should be made more rewarding and appealing not only to farmers but also to all actors in the market value chain, including consumers and traders. Value addition on the products of traditional landraces could be important entry points for future policy.

To boost the maize grain and groundnut (and thereby promote the local varieties such as Gankata and Kadononga), there is a need to provide loan facilities to traders involved to address their liquidity problems and financial constraints. Improving market storage facilities would help to ensure constant supply throughout the year, take advantage of economies of scale benefits, and enhance the use and market values of traditional varieties of crops. Moreover, addressing problems of transportation costs and poor storage facilities can reduce the seasonal variations in both supply and prices of the commodities. This will also induce farmers to produce more to meet the increased demand. Since most consumers prefer Gankata and Kadononga, and since there is no discrimination against local maize and groundnut varieties in the urban markets, these measures would ensure the increased availability of local varieties of these crops in the agricultural production systems of Zambia.

The study found non-existence of trade discrimination between the hybrids or improved varieties and local varieties of the crops studied. This is pertinent for the continued existence of the local crop varieties. However, this does not ensure that all potentially important traditional varieties are maintained. Some important varieties can be left out from the market value chain. There is a chance that some traditional varieties important for the future of agriculture could be crowded out. Identifying those varieties and taking measures to ensure their continuous survival is very important. Those traditional varieties which do not currently possess the relevant traits for consumers should be targeted for *in situ*/on-farm and *ex situ* conservation.

Notes

1 At the time of the survey, US$1 = 4000 Zambian Kwacha, the local currency.
2 'Samp' is pounded maize grits without husks that is boiled and eaten with the addition of either sugar, groundnut, salt or milk, or any combination of these.
3 Some small variations may exist which could not be attributed to differences between local and hybrid maize prices.

References

Bagachwa, W. S. D. (1994) 'Choice of technology in industry', in L. Rubey (ed.) 'Maize milling market reform and urban food security: the case of Zimbabwe', Working Paper AEE 4/94, Department of Agricultural Economics and Extension, Harare, Zimbabwe

CSO (Central Statistical Office) (2000) *National Census Report*, Central Statistical Office, Lusaka, Zambia

Edmeades, S., Smale, M., Renkow, M. and Phaneuf, D. (2004) 'Variety demand within the framework of an agricultural household model with attributes: the case of bananas in Uganda', EPTD Discussion Paper no 125, Washington, DC

GRPI (2006) *Promoting Incorporation of TraditionalVarieties and Breeds in Local Production Systems of Zambia:A GRPI Report*, University of Zambia, Lusaka, Zambia

Gumbo, M. (1986) *Maize Agronomy in Zambia*, Mount Makulu Research Station, Lusaka, Zambia

Hara, R. (2008) 'Factors affecting the adoption of hybrid maize varieties in Zambia's Chongwe district', BSc thesis, University of Zambia, Lusaka, Zambia

Hassan, C. (2005) 'The hammer milling sector after market reforms in Lusaka urban', BSc thesis, University of Zambia, Lusaka, Zambia

Howard, J. A. and Chitalu, G. M. (2000) 'Report on impact of investments in maize research and dissemination in Zambia', University of Zambia, Lusaka, Zambia

Kuntashula, E. (1999) 'Maize market reform: the position of the Lusaka urban milling industry and implications on household access to food in Lusaka, Zambia', MSc thesis, University of Zimbabwe, Harare, Zimbabwe

Mafuru, J., Sikani, F. and Bwanu, B. (1999) 'Adoption of maize production technologies in the lake zone in Tanzania', unpublished report, Tanzania

Rubey, L. and Masters, W. (1994) 'Unscrupulous traders, zero-sum games and other myths about grain market reform in Zimbabwe', University of Zimbabwe working paper no. AEE/1/94, Department of Agricultural Economics, University of Zimbabwe, Harare, Zimbabwe

Smale, M., Bellon, M. and Aguirre Gomez, J. (2001) 'Maize diversity, variety attributes and farmers choices in southeastern Guanajuato, Mexico', *Economic Development and Cultural Change*, vol 50, no 1, pp201–225

Wale, E., Mburu, J., Holm-Müller, K. and Zeller, M. (2005) 'Economic analysis of farmers' preferences for coffee variety attributes: lessons for on-farm conservation and technology adoption in Ethiopia', *Quarterly Journal of International Agriculture*, vol 44, no 2, pp121–139

Wale, E. and Yalew, A. (2007) 'Farmers' variety attribute preferences: implications for breeding priority setting and agricultural extension policy in Ethiopia', *African Development Review*, vol 19, no 2, pp379–396

Chapter 6

Commercialization and Market Linkages for Promoting the Use of Local Rice Varieties: A Nepalese Case Study

J. C. Gautam and Krishna Prasad Pant

Summary

Commercialization and marketing of traditional crop varieties (referred to as landraces) and their products is one of the major strategies to address conservation and sustainable use of crop genetic resources. The major policies related to commercialization of the genetic resources are designing mechanisms to:

- support and promote industries in using the genetic resources;
- promote seed development industries (including biotechnology);
- create gene markets, and encourage resource and credit flow; and
- encourage farmers and small entrepreneurs to diversify products from traditional varieties of crops.

This case study provides relevant information that will contribute towards developing a policy framework for the commercialization of traditional crop varieties and their products in Nepal.

The concept of farm business income as a tool for economic analysis is used to measure the willingness of Nepalese farmers to continue growing traditional landraces in the long term, and thereby competing with improved varieties in terms of yield and income. The results have shown that while utilizing their own genetic resources, which have limited alternative uses, rice producers have managed to get some benefit and enhance the productivity of the landraces. Moreover, retaining traditional varieties by the traditional farmers may also be explained by such economic rationales that give better resonance and resilience to the family, immunity to higher market fluctuations and protection against natural disasters.

This case study also presents the prospects of commercialization and potential for promoting the marketability of underutilized landraces as well as their products. The results have shown that local people are the main consumers of the local products, though many foreigners have also been consuming these products. The study suggests possibilities of promoting value added enterprises for some of the selected traditional crop varieties. With the enhancement of the profitability of such enterprises, farmers will be willing to participate in maintaining the diversity of traditional rice varieties that need public intervention.

Introduction

Commercial use of genetic resources (GRs) is one of the major strategies of effective biodiversity conservation. The use of GRs and their products in commercial value chains can generate income and other non-income benefits for conservationists and providers of GRs. This approach will also help in designing cost- and benefit-sharing approaches for the conservation and use of the resources.

Commercialization of GRs and their products highly depends on their potential market values. The market values, in turn, depend on the magnitude of commercialization, on the genetic technology industry's willingness to pay for samples of GRs and on the revenues a single provider can earn.

The major policy-related issues for the commercialization of GRs are to design mechanisms to support and promote industries in using GRs, which will then help to promote seed development industries including biotechnology; to create gene markets; to ensure credit flow for promoting commercialization of GRs; and to encourage farmers and small entrepreneurs to diversify GRs. Landraces with socio-cultural and market-preferred traits are few in number but have the potential to be conserved on-farm (Rana et al, 2007).

Taking commercialization and marketing of traditional GRs and their products as policy options to address the conservation and sustainable use of GRs, the purpose of this case study is to generate relevant information that will contribute towards developing a policy framework for the commercialization of traditional local rice varieties in Nepal. The study will explain the economic behaviour of traditional rice farmers who cultivate and maintain GRs at costs not compensated by the ongoing market prices. Despite the failure of the market to reward their contribution, farmers have continued to cultivate traditional crop varieties as long as they can meet the costs with their own home-grown inputs.

The study further reports relevant policy implications for promoting the products of neglected and underutilized local traditional rice varieties. Such products are gradually but slowly entering the market value chain. Supporting the development of niche product enterprises and forming public-private partnerships for the conservation and utilization of traditional varieties can improve both farmers' livelihoods and conservation outcomes.

Methodology

The farm income analysis of local rice landraces was undertaken in the Begnas area of Kaski district. Fifteen rice-cultivating farm households were selected for this case study. The following relationships for the income and the cost valuation of local landraces were applied:

Net income (NPR/ropani)[1] = Gross income – Total cost

Farm business income   = Gross income – Cost of purchased inputs
(NPR/ropani)

(Eqn 6.1)

Estimated farm business income is the income to the farm family from the crop after deducting the out-of-pocket costs of purchased inputs from the total income. The costs of own (non-purchased) inputs (labour, home saved seeds, manure, etc.) were not included in the cost calculation. Thus, the farm business income only gives a proxy of income as a return to own inputs and family labour. The study implicitly assumed that most of the farmers' own resources had very little alternative uses under the subsistence agriculture. Though this assumption may not be fully correct, the approach still explains farmers' rationale for engaging themselves in the traditional crop-farming business despite net loss (in the strict financial sense) in the cultivation of the crop.

Commercialization and marketing of traditional crop varieties

Farmers, consumers and sellers in Kathmandu Valley (Kathmandu, Lalitpur and Bhaktpur) and entrepreneurs and consumers of Palpa (Tansen Municipality and Suburb Pokhrathok), mid-hill areas of Nepal and Butwal haat bazaar (a Terai market with mixed community of hill and Terai people) were surveyed to assess the prospects of commercialization and the potentials for promoting and marketing local rice varieties (and their products) grown by Nepalese. Prospects for promoting commercialization of underutilized local rice varieties were also explored. Respondents were randomly selected for the interviews.

Description of the study sites

The study area (Kaski district) is in the mid-hill region (800–1500m above sea level) of Nepal. The topography of the region consists of ancient lake and river terraces found on moderate to steep slopes. It experiences high rainfall (>3900mm/annum) with a warm temperate to subtropical climate. Mean daily minimum temperature of the coldest month is 7°C and the mean daily maximum of the hottest month is 30.5°C with monthly mean of 20.9°C (Sthapit et al, 1999; Rana et al, 2000).

The area is reported to be a hotspot in terms of crop diversity (Rijal et al, 1998). A total of 32 crops were reported to be grown. The major crops are rice, maize and finger millet. Rice, the major staple crop, is grown in different environments (lowland, irrigated land, partially irrigated land, rain-fed and upland). The total rice varieties maintained by the local farmers in this area are about 69 (Rijal et al, 1998), 63 of them local (Rana et al, 2007).

Sampling for cost of production study

Out of 50 rice-growing farmers sampled in the Begnas area, 15 rice growers were sub-sampled for the analysis of input uses and cost of production for rice. The rice farmers in the district are homogenous and farm-to-farm variation is very low. Most of the farmers are growing rice with similar sets of inputs. The purchase of inputs and sales of outputs are done in local markets and most of the farmers fetch similar prices for their products. Due to such homogeneity among the farm households, even a small proportion of the total households can represent their situation well. Moreover, for the study on the marketing of underutilized local crop species and their products, more than three dozen entrepreneurs, shopkeepers and departmental stores (of Kathmandu Valley, mid-hills town like Tansen and a Terai-located town like Butwal, both in the western region of the country) were interviewed with a structured questionnaire. The questionnaires covered the production, production costs and rice variety attribute preferences of the interviewees.

Input cost and income analysis

Table 6.1 reports input cost and income analysis of eight local varieties of rice. Net income was calculated by deducting the total costs from the gross income. Farm business income is obtained by deducting the costs of purchased inputs from gross income (see 'Methodology', above). As there is neither tax nor subsidy on farm income, the farm business income is the family income. The traditional varieties – Anga, Chotte, Local Mansuli and Mansara – were found to provide negative net incomes to the farmers. However, excluding Mansuli, they generated positive farm business income. For Anadhi, average farm business income (NPR

1808) was almost five times its net income (NPR 362). Jetho Budo ranked very high in net income (NPR 1393) and provided the second highest farm business income (NPR 2392). Bayarni Jhinuwa provided the highest average farm business income of NPR 2438, but its net income was almost half of the Jetho Budo. Farmers cultivating Bayarni Jhinuwa gained higher benefits by using more of their own inputs and less purchased inputs. Farm income of these farmers was higher than that of Jetho Budo farmers, though the latter earned more net income. Maintaining traditional varieties by traditional farmers might be partly explained by the better resonance and resilience to the family, immunity to higher market fluctuations and protection against natural hazards.

Jetho Budo and Anadhi resulted in better net income because they were in high demand and their market price was also higher than that of other varieties. Despite the net loss which farmers faced by growing some other landraces, their farm business income from their home-grown inputs (landraces) was still positive. Therefore, farmers grew rice landraces and made maximal use of their own inputs. If they did not use landraces, their own inputs had no or very little alternative uses. Thus, these traditional varieties enhanced the value of farmers' own resources. Further, these local varieties have their own cultural importance. For instance, Anadhi was used in special festivals for making 'latte' (rice made with large amount of ghee) because it absorbs ghee during cooking. Due to their unique characteristics, such rice varieties are highly demanded by consumers and farmers therefore continue to cultivate them. Farmers' rationale in maintaining traditional varieties need to be further examined so that effective partnership programmes (that can enhance the productive capacity of farmers' landraces, improve farm income and ensure the continuous survival of traditional rice varieties) can be developed.

Seed procurement, storage and sales of local rice varieties

The sources of seed, its acquisition and replacement systems differ from farm to farm. In Begnas, about 93 per cent of farmers retained their own rice seed for next year planting. About 5 per cent of farmers received seeds from neighbouring farmers and only about 2 per cent of farmers obtained seeds from development organizations (NGOs, cooperatives, seed-selling enterprise, etc.).

The farmers were asked where they sell seed paddy. About 73 per cent of them reported that they sold it to their neighbours, not for money but in exchange for paddy. The prices for seeds are not different as the rice for seed is not produced differently. In the previous year, only 20 per cent of the respondents sold seeds to the cooperatives and NGOs. The cooperatives and NGOs were reported to occasionally visit villages to buy seeds.

Table 6.1 *List of own and purchased inputs and derivatives of incomes from different local rice varieties*

Cultivar	Own inputs NPR/Ropani					Purchased inputs NPR/Ropani				Total cost	Gross income	Net income	Farm business income
	Seed	Farm yard manure	Labour	Bullock	Total cost of own inputs	Inorganic fertilizer + insecticide	Bullock	Labour	Total cost of purchased inputs				
Anadhi													
Mean	67.5	173.8	1117.5	87.50	1446.25	66.50	193.8	900.00	1160.25	2606.50	2968.75	362.25	1808.50
SD	9.57	176.8	810.69	105.0	926.5	81.93	229.5	483.05	473.72	1327.99	952.05	2264.84	1389.97
Anga													
Mean	55.0	200.0	900.00	155.00	1310.0	131.2		1645.83	1777.00	3087.00	2433.33	-653.67	656.33
SD	7.07	70.71	636.40	7.07	551.54	77.55		206.24	128.69	422.85	329.98	752.83	201.29
Bayarni Jhinuwa													
Mean	70.0	225.0	1140.0	264.0	1699.0			862.5	862.5	2561.5	3300.0	738.5	2437.5
SD	0.00	106.1	84.85	8.49	199.40			371.23	371.23	570.64	424.26	994.9	795.5
Chotte													
Mean	55.6	179.2	765.8	199.17	1199.7	118.8		1377.78	1457.0	2656.7	2594.4	-62.28	1137.4
SD	17.4	26.02	444.13	65.78	500.97	27.11		302.46	281.23	721.16	356.03	419.77	299.0
Ekle													
Mean	57.0	184.6	275.84	113.83	631.33	32.67	235.8	1059.83	1328.28	1959.62	2759.88	800.26	1431.59
SD	14.7	76.85	218.46	68.91	220.72	56.58	77.76	335.37	406.93	551.12	650.37	449.69	316.36
Jetho Budo													
Mean	70.6	306.3	515.89	105.50	998.26	131.0	222.1	1680.11	2033.21	3031.47	4424.83	1393.36	2391.62
SD	11.5	209.5	78.07	49.38	207.42	227.3	99.74	822.61	954.59	1065.31	661.19	1372.48	1333.19
Local Mansuli													
Mean	58.9	290.4	375.0	0.00	724.29	387.2	342.0	2562.50	3291.64	4015.93	3107.14	-908.79	-184.50
SD	22.7	119.7	106.07	0.00	248.5	116.4	41.67	441.94	516.69	765.19	151.52	916.71	668.22
Mansara													
Mean	54.0	182.5	820.50	162.0	1219.0	54.8	135.0	1282.5	1472.3	2691.3	2360.0	-331.3	887.7
SD	8.22	72.67	402.62	147.89	582.99	122.5	186.8	495.57	657.82	816.39	357.77	979.47	753.99

Notes: 1 hectare = 19.66 ropani; US$1 = NPR 64.45 on 6 April 2008; SD = Standard deviation.
Source: 2006 household survey, Nepal

Table 6.2 *Difference in price of paddy and milled rice*

SN	Landraces	Price of paddy (NPR/quintal)			Price of milled rice (NPR/quintal)		
		Right after harvest	After 12 months of harvest	% of price difference	Right after harvest	After 12 months of harvest	% of price difference
1	Jetho Budo	1843.0	2158.8	17.1	3996.03	4133.8	3.5
2	Anadhi	1722.4	2204.8	28.0	3720.4	3766.4	1.2
3	Local Mansuli	1056.4	1378.0	30.4	2618.1	3307.1	26.3

Note: Paddy to rice conversion percentage (milling recovery percentage) is approximately 50–60% depending on the variety.
Source: 2006 household survey, Nepal

Price variation across rice landraces

The price of rice landraces was, as expected, found to be different across seasons. The price of rice was found to be lower shortly after the harvest than after 12 months of harvesting. Consumers preferred one-year-old rice (for its desirable consumption traits) to newly harvested rice which is part of the reason for the price variation. During the harvesting season, both the one-year-old rice and the newly harvested rice are sold side by side but at different prices. The percentage difference in prices during harvest time and the later period was higher for paddy than for the milled rice. The difference for paddy ranged from 17 per cent to 30 per cent whereas in the case of milled rice the difference ranged from 1.2 per cent to 26 per cent. Table 6.2 reports the price differences during harvest and 12 months after the harvest.

Traditional farmers depend on the farm production for their food security and income. To achieve their household food security, the farmers attempted to grow high-yielding varieties, with the consequence that farmers more often chose to cultivate modern varieties, pushing the low-yielding landraces to the verge of extinction. At the same time, some better-off farmers wanted to meet their socio-cultural needs by growing landraces with unique properties. Farmers always seemed to face trade-offs between high yield and preserving their unique culture and taste. Commercialization of the landraces and increased income from them seemed to attract farmers towards the landraces, thereby contributing to their conservation.

Marketing channel

A marketing channel is the path through which the commodity flows via different traders from producers to consumers. As far as the marketing channel of rice seeds is concerned, farmers can be considered producers and also consumers. In Nepal and in many other developing countries, seed distribution systems,

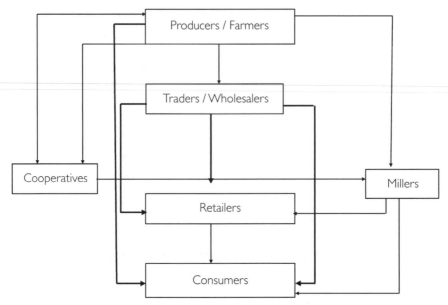

Figure 6.1 *Marketing channel of local rice cultivars*

Note: Producers are also consumers. Sometimes, paddy seeds are sold by cooperatives to producer farmers. Traditional systems of milling also prevail in some areas.
Source: Based on the discussions and interviews held at Begnas area of Kaski District

particularly for local producers, are mostly based on barter through exchange among neighbouring farmers.

Agricultural products were reported to pass through different functionaries/channels involving various economic actors before reaching the final consumers. In the case of the present study area, such actors include collectors/vendors, processors, group suppliers, wholesalers and retailers. In general, producers sold their paddy to traders or local millers. In some cases, the cooperatives were working in seed marketing. Cooperatives were reported to buy seeds from farmers and resell to other farmers. The general system of marketing of local paddy cultivars in Nepal is depicted in Figure 6.1.

Commercialization and marketing of local crop varieties

In the framework of GRPI-Nepal, a market survey on underutilized local crop species was conducted in Kathmandu Valley and Tansen, a less urbanized town in the mid-hills. The survey was done for various local crop species that also included neglected and underutilized local species such as 'maseura', 'buckwheat flour', 'finger millet flour', 'gundruk', 'soybean', 'horsegram', 'perilla', 'sesame', 'rice bean', 'cowpea', 'bhang', 'blackgram', 'timur', 'jimmu', and 'ash gourd' (ABTRACO, 2006).

The number of consumers buying local products from local stores and whole-salers ranged from 5 to 500 per week in Kathmandu Valley. These products were bought for their medicinal value and because foreigners were attracted towards these indigenous products. Due to availability of market demand, most of the stores expressed an interest to deal with these products. They also noted that export markets of such products are India, Japan, Israel and the USA.

The prospects for promoting value addition enterprises

According to the retailers, local people were the main consumers of the local rice products. At the same time, many foreigners were also reported to have been consuming local products. The study suggests possibilities of promoting value addition enterprises for some of the selected local crop varieties. Obviously, with enhancement of such enterprises, farmers might be willing to participate in maintaining the diversity of the species.

Price margins for sellers

Price margin is defined as the difference between the price paid by the consumers and the buying price of the sellers. In this study, the price margin of sellers was calculated on the basis of the purchasing and selling prices of the sellers.

In the case of certain items such as maseura, perilla, sesame and rice bean, the wholesale and retail prices were reported to be the same. This is mainly because in many cases the retailers are also the producing entrepreneurs who are also the sellers to other sellers such as department stores. This shows a small-channel and emerging nature of the markets of these species.

The price margin for selected department stores often varied, i.e. NPR 2 (10 per cent price margin) for gundruk to NPR10 (30 per cent) for soybean per kg. In the case of retailers, it varied from NPR 2 (5 per cent) to NPR 20 (75 per cent) for perilla, cowpea, timur and jimmu. The other secondary sellers also share the margin (the difference between selling and buying price) from NPR 5 (33 per cent) to NPR 20 (40 per cent). Prima facie, there were several possibilities of promoting the marketing of local varieties of crops and their products. However, the pricing structure and analysis of the enterprise establishment need further research, incentives and policy measures for promising public-private partner-ships to be devised.

Conclusions and implications for policy

Promoting the commercial use of traditional varieties of crops and marketing their products is one of the important approaches for utilizing, enhancing and conserving crop genetic resources. This can support the sustainable development of agriculture in two ways:

1   by improving farmers' incomes/livelihoods; and
2   by ensuring the conservation of these resources on-farm.

To augment these objectives, farmers should be provided with appropriate technologies that help reduce the production cost to get higher benefit from the local rice varieties. Small industries or micro-enterprises based on products of traditional crop varieties (landraces) should be established on a participatory basis.

Net incomes for some traditional varieties were found to be negative; they were less preferred and grown on small areas. However, the growers of these and most of the other local varieties of rice earn positive farm business income. By utilizing their own farm-grown resources, which often do not have other alternative markets, rice producers managed to gain benefit and enhance the productivity of underutilized traditional varieties. This is why farmers are growing rice landraces in spite of the net loss in net farm income. Moreover, retaining traditional varieties by traditional farmers may also be explained by their better resonance, better resilience to local households, increased immunity to higher market fluctuations, and better protection against vagaries of nature – factors contributing to the livelihoods and household economics of local farmers.

The local species that are not producing positive net farm income may be discarded by farmers in the long term, especially when access to improved varieties and other inputs increases. However, varieties with important traits should be conserved for breeding purposes through additional public efforts, which will be more successful if they are linked to the livelihoods of rural communities.

Scented rice varieties like Jetho Budo have high values for consumers. Thus, a programme of development of aromatic rice needs to be implemented with the active participation of farmers. A consolidated network of production, processing, marketing and consumption of aromatic rice could be launched. Expansion of market linkages with the promotion of enterprises for value addition and marketing activities for local varieties (plus their products) should be carried out (ABTRACO, 2007). The values and importance of the products of local rice varieties in generating incomes and employment for the local entrepreneurs should be recognized. Further research on nutritional values, product design, processing and attractive packaging of the local products needs to be undertaken.[2]

Notes

1   NPR is Nepalese rupee. Ropani is a unit of land which is approximately $507.8m^2$.
2   This study was sponsored by Bioversity International and was conducted by
     ABTRACO. The authors duly acknowledge the support of Bioversity and the
     comments and inputs of all editors and the GRPI team, especially the unlimited
     technical and editorial inputs provided by Dr Edilegnaw Wale.

References

ABTRACO (2006) '*Analysis of National Policy for Promoting Underutilized Species*', Agri-
     Business and Trade Promotion Multipurpose Cooperative, Nepal
ABTRACO (2007) '*Commercialization and Market Linkages for Promoting the Use of Plant
     Genetic Resources*', A report submitted to Bioversity International, SSA, Nairobi, Kenya
Rana, R. B., Garforth, C., Sthapit, B. and Jarvis, D. (2007) 'Influence of socio-economic
     and cultural factors in rice varietal diversity management on-farm in Nepal',
     Agriculture and Human Values, vol 24, no 4, pp461–472
Rana, R. B., Rijal, D. K., Gauchan, D., Sthapit, B., Subedi, A., Upadhayay, M. P., Pandey,
     Y. R. and Jarvis, D. I. (2000) '*In situ Crop Conservation: Findings of Agro-ecological, Crop
     Diversity and Socioeconomic Baseline Survey of Begnas Ecosite Kaski, Nepal*', NP working
     paper no 2/2000, NARC/LI-BIRD, Nepal/IPGRI, Rome
Rijal, D. K., Rana, R. B., Sherchand, K. K., Sthapit, B., Pandey, Y. R., Adhikari, N.,
     Kadayat, K. B., Gauchan, Y. P., Chaudary, P., Poudel, C. L., Gupta, S. R. and Tiwari, P.
     R. (1998) '*Strengthening the Scientific Basis of In situ Conservation of Agro-biodiversity:
     Finding of Site Selection, Kaski, Nepal*', NP working paper no 2/1998, NARC/LI-BIRD,
     Nepal/IPGRI, Rome
Sthapit, B., Upadhyay, M. and Subedi, A. (1999) *A Scientific Basis of In situ Conservation
     of Agrodiversity On-farm: Nepal's Contribution to Global Project*, NP working paper no
     1/99, NARC/LI-BIRD, Nepal/IPGRI, Rome

Part 4

Conclusions and Outlook

Chapter 7

Findings, Conclusions, Implications and Outlook

Edilegnaw Wale, Kerstin K. Zander and Adam G. Drucker

In this concluding chapter, we provide a synthesis of the economic methods used in the case studies, a summary of the findings and associated implications, as well as suggesting some directions for future research.

Recapping the issues and the economic methods

This book has sought to document a variety of economic issues relating to agro-biodiversity policy, with particular reference to the conservation of plant genetic resources used in food production. Different GRPI project country case studies (Ethiopia, Nepal and Zambia) were used to address various economic research questions identified through a participatory process. The information obtained can be used to better account for, and potentially mitigate, any negative externalities that may arise from development interventions, such as, *inter alia*, the introduction of improved crop varieties and improved market access.

In this chapter we provide a summary of the main findings and implications for farm management and policy decisions related to the *in situ* conservation of plant genetic resources used in food production. Many of these resources have a range of important non-market values but have low financial profitability and/or are underutilized. Consequently, the case studies presented in this book require and use a variety of economic methods to account for these diverse values. The applied methods include choice experiments, hedonic pricing, variety attribute preference ranking, contingent valuation and farm business analysis. The empirical analyses focused on a range of issues related to farmers, consumers and traders' preferences for traits and varieties; farmers' perceptions of the livelihood impacts of the replacement and loss of traditional crop varieties; and the commercialization/marketing of, and value chain development for, traditional crop variety products.

By presenting the various case studies, the book not only aims to strengthen the support of economic analysis for a better understanding of on-farm values of crop genetic resources, but also demonstrates the market and consumer values associated with crop landraces (e.g. taste). Emphasizing these kinds of values can support the generation of sustainable sources of conservation funding, for example, by facilitating the provision of niche products from landraces for which consumers are willing to pay a premium. In the best case scenarios, this premium acts as a direct contribution to farmers' income and may favourably influence farmer decisions regarding switching to improved varieties, continuing with landraces, or using both.

The partners and stakeholders engaged with the GRPI project have identified the issues and found the subsequent synthesis of the economics work useful in supporting their policy development processes. This demand-driven work was carried out with the expectation that such an approach will facilitate the uptake of the results by rural development policy-makers in the respective GRPI countries. In this context, the book will contribute to showing how addressing economic questions can feed into the genetic resources policy process. The overall findings are relevant not only to the GRPI countries involved in the study but also to other countries concerned with the sustainable utilization of such resources. Above all, the book will have achieved its objectives if it can illustrate how genetic resources issues can be integrated into development interventions to address potential policy trade-offs and if the issues addressed are picked up by the decision-makers in the respective GRPI countries.

As demonstrated in this book, different crops have different values to different actors in the crop production and distribution value chain. Naturally, different economic analysis methods are suited to different types of genetic resources and different types of policy issues. The choice of specific method, for instance, depends to a great extent on the type of values being assessed (e.g. monetary, non-monetary, current, future, private, public, etc.), the economic questions we wish to answer, and the availability and potential sources of data for analysis, as well as the time-frame and budget available, as some approaches are more complex/expensive than others.

Summary of the main findings

The findings have shown that the cultivation of crop varieties (teff and sorghum in Ethiopia; rice in Nepal; and maize and groundnut in Zambia) depends not only on farmers' perceptions, preferences and the utility they derive from the cultivated crops (as shown in Chapters 2, 3, 4 and 6) but also on their market prices and consumers' demand for their products (Chapters 3, 5 and 6).

Teff and sorghum variety traits in Ethiopia related to yield, environmental adaptability and yield stability, are valued in Chapter 2 using a choice experiment approach. Producers' willingness-to-pay (WTP) estimates were derived for both crops considering the most important and relevant traits (yield, price, environmental adaptability and yield stability). For both sorghum and teff, the highest WTP was associated with environmental adaptability (256 Birr and 516 Birr per quintal, respectively). Producers' WTP for the yield-stability trait for sorghum and teff is estimated to be 229 Birr and 360 Birr, respectively. By contrast, WTP for yield per se was much lower for both crops (14 Birr for sorghum and 25 birr for teff). Such a result may be explained by sale price variability and poor market conditions during better harvests. Farmers' differential valuation of crop variety attributes were shown to be associated with differences between farm households with respect to their endowments, constraints and their level of developmental integration (in terms of basic infrastructure, access to markets and agricultural extension).

The case study of rice in Nepal (Chapter 3) had a similar goal of assessing the value of specific crop variety traits. Combining hedonic pricing and contingent valuation methods, both consumers' and farmers' WTP for traits of different rice varieties were considered. Aroma and taste traits were found to have a national-level value of NPR 11 billion ($148.6m) and NPR 2 billion ($27m) per annum, respectively. At the national level, farmers value a combination of high-yielding and aromatic traits at nearly NPR 1 billion ($13.5m) and for traits related to expansion during cooking and disease resistance at over NPR 1 billion ($13.5m). As might be expected, compared to consumers, farmers also derive higher utility from the yield and disease-resistance traits, as these two traits play an important role in income generation.

The estimations of value included the direct use value from consumption by the households and option values generated by an individual's WTP to protect the rice landraces for the future use in rice breeding. The value given by the farmers to use seeds of an aromatic landrace is derived from the value given by the consumers. Other direct use values related to the ecosystem functions of paddy field and option values were also considered. However, non-use values were not taken into account in this study and the authors concluded that the values quantified were consequently lower-bound estimates.

Ethiopian farmers' perception of the loss/replacement of traditional crop varieties and the impact of the loss to their livelihoods is analysed in Chapter 4. That chapter deals with a subject that is left to scientists to analyse, debate and make recommendations. Farmers are found to be knowledgeable about the process of creolization (i.e. the adaptation of improved varieties to local agro-ecological dynamics) and the role of local varieties (as inputs or raw materials) in improved variety development. According to most farmers, the composition of the crop varieties they grow in their villages has indeed changed during the last five years as a result of the replacement of traditional varieties by improved ones, resulting in a decrease in the probability of finding traditional varieties on farmers' fields. Farmers have reported that the traditional varieties are losing their desirable traits and becoming incompatible with the poverty of the soil.

Results of a logistic regression analysis of that chapter further revealed that such perceptions are held by farmers who are more experienced, are better networked, have a better chance to get improved seeds and have greater preference for crop varieties that can fetch higher prices. By contrast, farmers who have higher preferences for yield stability and early maturity traits (and hence are typically dependent on traditional varieties of crops) do not perceive such replacement to be occurring to the same degree, arguing instead in favour of the adaptive traits of local varieties. These farmers are typically net sellers of agricultural products. Livelihood impacts tend to be felt by farmers in terms of the perceived (in)accessibility to improved seeds and the high cost of improved seeds, and thereby lower farm income. Farmers whose livelihoods are affected by the loss of traditional varieties are also missing the health and nutrition benefits of traditional varieties of crops. The results suggest that these farmers face frequent food security problems and have a greater preference for the yield-stability trait. On the contrary, farmers who do not see their livelihoods being affected by the loss of traditional varieties have a lot of trust in the superiority of improved seeds with which they have better experience.

By recognizing the links between consumption, production and on-farm utilization along the market value chain, and ranking variety attribute preferences, differences in attributes influencing decisions to purchase maize (Gankata) and groundnut (Kadononga) in Zambia are explored in Chapter 5. The chapter typically deals with the preferences of urban consumers and traders.

The findings show that as a result of being a staple food, quantity-specific attributes (such as grain size and kernel density) are very important for maize. Facing poverty and food insecurity, consumers prefer to maximize the quantity of maize for a given price and are not willing to pay a premium for better quality maize varieties such as those with better taste. By contrast, when consumers make decisions to buy groundnut (a non-staple), quality attributes (e.g. taste) were perceived more important. Such preferences explain the widespread existence of the most common local maize (Gankata) and groundnut (Kadononga) varieties, as these possess preferred attributes. On the other hand, the disappearance of other local varieties could be associated with their absence of the desired traits in those varieties. The income status of consumers, proximity to markets, purpose of purchasing maize (as a staple: nshima) and preferences for local varieties over improved ones are the most important factors explaining the quantity of maize purchased. The direct and significant impact of household characteristics (such as growth in family size) on quantities of maize bought also demonstrates the importance to consumers of maize availability over quality.

The chapter demonstrates interesting contrasts between maize and groundnut. With groundnut grain, the traits that received more preference were food taste and size of grain, which are equally more important than oil content, kernel density and grain colour. Market price, easiness to peel and the number of pods are the least important traits for groundnut. Size of groundnut was also not very important. According to most consumers, Kadononga, a famous traditional groundnut variety, possesses most of the desirable attributes of groundnut, which again explains its fame and survival for so long.

Chapter 6 has argued that the commercial use of genetic resources, which depends on the market value of their products, is one of the major strategies for achieving sustainable agro-diversity conservation. Where such commercial potential exists, there remains the possibility of protecting and utilizing neglected and underutilized local crop varieties through public-private partnerships supporting the development of diversified and niche local product markets. In support of such an argument, aspects of the commercialization of local rice varieties in Nepal and their contribution to farmers' livelihoods are investigated using a gross margin approach.

The results show that gross margins vary across different types of traditional rice varieties, ranging from those which generate negative margins (Bayarni Jhinuwa, Anga, Chotte, Mansuli and Gurdi) to those which are very profitable (e.g. Jetho Budo). Although the landraces with negative margins tend to be allocated less land than high value cultivars, they continue to be cultivated as their output value still outweighs their purchased input costs. Farmers cultivating varieties like Bayarni Jhinuwa are benefiting in terms of employing their own inputs and less of purchased inputs. They grow these rice landraces and make use of their own inputs which have lower or zero shadow price and very little alternative uses. By employing their own resources which otherwise could not have been used productively, farmers manage to generate some subsistence income. To cope with cash shortages, they have to reduce the purchased inputs and use more of their own inputs. That is one of the reasons for traditional farmers to retain traditional varieties despite very low incomes. Additional unquantified economic benefits include better employment opportunities for the family labour, buffering against market fluctuations, insurance against varying climatic impacts, early maturity and cultural values.

Furthermore, consumers were found to be willing to pay a premium for scented/aromatic rice. This suggests that an aromatic variety development programme could be implemented with the participation of farmers and other actors in the marketing chain.

Conclusions and implications for genetic resources policy

The various topics empirically examined in this book reveal the importance of understanding the link between farmers' livelihoods (strategies and incomes), consumers' preferences and crop diversity outcomes. In other words, there can be no sustainable conservation if farmers' concerns/preferences/incomes are not linked with crop variety traits and the marketability of traditional varieties of crops. As far as agro-biodiversity is concerned, conservation for maintenance sake is unlikely to succeed. There needs to be a paradigm shift in the agro-biodiversity arena. Conservation (with a human and crop diversity face) has to be the policy objective. This can be achieved through, *inter alia*, enhancing the production of local varieties, value addition and the commercialization of traditional variety products. As a complement to *ex situ* conservation strategies, *in situ* conservation creates untapped opportunities to link rural development interventions with conservation of these resources.

Sound conservation policy needs to be based on a better understanding of the types of varieties preferred by different farm household types (Chapter 2). This permits the identification of varieties conserved *de facto* and those that need additional incentives to ensure their continued survival. For instance, *de facto* conservation of environmentally adaptable sorghum varieties implies that there is no need to design incentives to maintain these varieties. By contrast, for those varieties not maintained *de facto*, strategies and policies need to be designed to ensure their conservation (Chapters 2 and 3). In particular, this will be necessary where a certain variety trait is unique (i.e. contributing significantly to overall diversity) but current demand for that trait is low, leading to declining use and an increasing extinction threat. Of course, such strategies would have to be responsive to dynamics in preferences (of both consumers and farmers), agro-ecological factors (e.g. climate change, land degradation, drought and desertification), opportunities and institutions (e.g. new markets, new crop enterprises, new varieties, etc.). However, it may nonetheless be the case that conservation costs may be relatively small compared to the value of the unique traits conserved (Chapter 3). That is why this chapter concludes that protecting each of the preferred and non-preferred rice variety traits adds value to society.

Gaining an improved understanding of variety choice and the preference for adaptive traits informs not only conservation policy but also future breeding activities. Given that poorer teff and sorghum farmers highly value environmental adaptability and yield stability (Chapter 2), future breeding programmes should incorporate such traits into their agenda rather than tending to concentrate on productive traits. To target and address variety demands of poorer farm households, the priority variety attributes are the environmental adaptability and yield stability of both teff and sorghum varieties, not yield/productivity which is often given more attention by breeding programmes. Part of the explanation for farmers' relative lack of interest in yield is the decline of prices during good harvests because of the lack of effective market demand to absorb the extra production. In terms of opportunity cost compensation (Wale, 2008), poorer farm households most affected by losing varieties with better yield stability and environmental adaptability will require equivalent compensation if the policy entails denying them the use of those varieties. Where farmers continue to face opportunity costs by growing traditional varieties, it will be essential to compensate poor farmers for maintaining low productive landraces since, in the long term, they cannot afford to bear the opportunity costs of on-farm conservation. To avert potential loss, conservation agents have to target those varieties that have demand from neither farmers nor consumers and pursue income support through compensatory measures (Chapters 2, 3 and 6).

However, consumers and farmers differ in their variety attribute preferences (Chapters 3 and 5). The cultivation of crop varieties depends not only on farmers' perceptions and their utility from the traditional varieties of cultivated crops (as shown in Chapters 2, 3, 4 and 6) but also on consumers' demand for those varieties (Chapters 3 and 5). While consumers place more value on consumption traits (such as aroma), farmers place more value on production traits. In Zambia (Chapter 5), for example, the empirical results have shown that future breeding efforts should target quantitative traits (like kernel density and grain size) for staple food crops (like maize) and quality traits (like food taste) for non-staple food crops (like groundnut). Thus, breeding objectives need to account for both sets of demands in order to ensure that farmers will have access to varieties demanded by consumers.

Where demand for varieties with outstanding quality traits is low, *de facto* conservation is unlikely to happen because it is financially not attractive for farmers to cultivate these varieties. One strategy for conserving such varieties is to identify alternative markets (local or international markets) in order to improve marketing channels and generate a higher volume demand for these varieties and/or their products. In this context, the preference of consumers for local varieties over hybrid or improved ones (for cultural and taste reasons) can be taken as an opportunity to improve farmers' incomes and conservation outcomes. Campaigns, seed fairs and commercialization efforts that build on and benefit from the preferred attributes of local varieties may also have a role to play in this context, leading to an increase in urban demand (for example, for local maize and groundnut) and thereby increasing their production, use and on-farm conservation.

Expansion of market linkages, together with the promotion of enterprises for value addition, for traditional varieties and their products needs to be carefully considered. The process can lead to the crowding out of some traditional varieties important for the future of agriculture. In this context, marketing studies and value-chain analyses can play an important role in terms of identifying the missing links and constraints of relevance to the promotion of enterprises based on products made using traditional varieties. Those traditional varieties that do not currently possess the relevant traits for farmers/consumers should be identified and targeted for conservation. Technologies that can enhance the comparative advantages of local varieties and result in value addition on their products can be important entry points to ensure the survival of such varieties perceived to be useful by farmers, consumers and traders.

Another area of conservation policy is related to the need for facilitating farmers' access to appropriate technologies that can help reduce local variety production costs and increase the benefits of their marketing/commercialization. This might include research and development (R&D) activities that can address the nutritional values, product design, processing and attractive packaging of the products of traditional varieties. These actions can support the three pillars of sustainable development (discussed in Chapter 1). Appropriate recognition should also be given to the synergic role of *in situ* conservation to the rural economy in generating employment and incomes for local entrepreneurs and the additional multiplier effects to the economy at large through the development of small-scale agribusinesses. Incentives and policy measures promoting public-private partnerships should also be considered in this context.

The last area of agro-biodiversity conservation policy is related to the importance of linking the dissemination of improved varieties with on-farm conservation initiatives for traditional varieties (Chapter 4). Where the former displace traditional varieties to the point of extinction and impose costs on society by undermining overall diversity, an externality is generated. Presenting improved varieties to farmers as a panacea can lead to the irreversible loss of traditional varieties, even in cases where the improved varieties eventually prove to be inferior. Given that trade-offs exist between access to improved seeds and the survival of local varieties, agricultural extension programmes that involve the introduction of new (improved) varieties need to ensure that this is done in a participatory manner that engages farmers, and integrates their views and perceptions as to what it means to crop diversity and their living conditions. The results also suggest that socio-economic changes and agro-ecological dynamics in smallholder farming often work against the comparative advantages of traditional varieties of crops, especially when those varieties retire and become incompatible with the changes.

Furthermore, where agricultural extension and dissemination of improved varieties may seriously threaten the continued use of traditional varieties, those farmers who are more likely to continue to value the traditional varieties (e.g. because they are isolated from input markets or they prefer adaptive traits or they consider taste and socio-cultural aspects to be of particular importance) are those who should be targeted to participate in on-farm conservation initiatives to facilitate the attainment of wider (e.g. national) diversity conservation programme goals. Support mechanisms may include facilitating the exchange of information among traditional variety-growing farmers (i.e. pro-diversity farmers/communities) so as to share (with other farmers) their knowledge/experience and create synergic collective action outcomes. However, such pro-diversity farmers should not become forced to retain traditional agriculture practices simply to maintain crop diversity for the public good. They should either get equivalent compensation for their public contribution or the productivity of their traditional varieties needs to be enhanced.

Outlook: the road ahead for economics and genetic resources policy

The case studies in this book have shown that farmers' preferences, values, responses to new genetic technologies, and potential to add value and market traditional crop varieties need to be integrated into genetic resources policy in order to facilitate both conservation and the enhancement of farmers' livelihoods. In this final section, we seek to highlight some of the main issues and unanswered questions that need to be tackled to better integrate agro-biodiversity issues in future rural development policy decisions.

The majority of crop genetic resources are cultivated in developing countries where markets tend to be segmented, volatile, unreliable, risky and poorly connected to other businesses. Farmers have limited alternative distribution channels for their local products. Global changes are also occurring that are likely to further influence the status of crop genetic resources in the foreseeable future. These changes can be institutional, market-related and environmental.

Institutional and market changes

Globalization is likely to play a key role in determining the future ability of farmers to compete in crop and seed markets. For example, changing food market conditions (e.g. commodity price increases and the growth of supermarkets) may affect landrace use and conservation. This has particular implications for farmers dependent on traditional varieties and the incentives they face to sustainably use threatened and underutilized crop varieties. Questions to be raised include, *inter alia*: whether and how rising food prices and the emerging competition between food and biofuels production will affect the status of local crop varieties; how might price changes and alternative uses of traditional food crops affect poor farmers; and how the cultivation of local varieties will be affected by input cost (e.g. transport, fertilizer) changes.

Institutional economic analysis, market chain analysis and exploring value addition opportunities can support improved understanding of: the income share of all actors involved in the market chain; the functions performed; the value added at each step of the chain; and the opportunities and constraints related to the flow of a particular variety and of potential entry points into the market (Wale, 2006). In some cases, such analyses will reveal opportunities for niche market development as an avenue towards promoting conservation through sustainable use.

To build on the work done as part of GRPI in identifying the importance of the link between local variety production, value chains and marketing networks, there is a need to extend these studies to understand the factors that may permit improved commercial exploitation (e.g. how local variety competitiveness can be increased; the threshold market size for commercially marketing a local variety; and how niche markets can be identified and exploited). Such studies would help in distinguishing between genetic resources that can be conserved by self-sustaining market development and those that will have to be treated as a public good requiring other public interventions for conservation. In both cases, improved understanding is needed with regard to the ways and means of increasing the comparative advantage of traditional varieties of underutilized crops through the use of better production, processing and marketing methods/technologies.

Facilitating conservation through the use of compensation or support payments could also be explored within the context of 'payment for environmental services' (PES). PES schemes have so far tended to focus on carbon sequestration and storage; non-domesticated biodiversity protection; watershed protection; and protection of landscape aesthetics. A review of the PES literature (see Mayrand and Paquin, 2004; Dasgupta et al, 2008; Wunder, 2008), covering hundreds of PES and PES-type schemes, reveals that there has been almost no explicit consideration of PES in the context of agro-biodiversity. The ability of 'payment for agro-biodiversity conservation services' (PACS) schemes to permit the 'capture' of public conservation values at the farmer level and thereby create incentives for the conservation of agro-biodiversity and supporting poverty allevi-ation, therefore appears to be worth exploring. However, implementing such a policy development strategy would require an assessment of the degree to which generic PES scheme opportunities and constraints might apply to PACS.

Underutilized crops are locally abundant, globally rare, little-known scientifi-cally but known in-depth by farmers and local community; most importantly their use is limited compared to their potential (Gruère et al, 2009). If countries continue to depend on a handful of crops, markets will face price shocks subject to the supply fluctuations of these few crops. Recent food price increases can in part be attributed to our dependence on a handful of food crops. Where change and interventions can support increased multiple variety and traditional crop competitiveness, increasing the food crop portfolio (by exploring the potential of traditional food crops) may lead to the generation of hitherto unexploited economic opportunities (in terms of new products, new job opportunities, new income sources, price stabilization and diverse food choices). If such benefits trickle down to disadvantaged smallholder farmers, not only will farmers' incomes be enhanced but also on-farm conservation of the diversity of the crops is more likely to be ensured. Significant R&D work needs to be undertaken on the ways and means of increasing the comparative advantage of traditional varieties of underutilized crops through use of better production, processing and marketing methods/technologies.

Environmental changes

Concerning environmental and agro-ecological changes, there is a strong need to analyse their effect on farmers' capability to continue with the cultivation of local crops and to consider farmers' adaptation and coping strategies when facing natural disasters (in particular, droughts and floods) and agro-ecological changes (climate change and land degradation). The impact of such changes on the comparative advantages of cultivation of local varieties needs to be explored urgently. If research shows that such changes are having a negative impact on agro-biodiversity, mitigation measures will have to be designed.

Methodological developments

Further methodological advances within the field of genetic resource valuation will also continue to be important. For example, there is potential to extend recent applications of choice modelling (Zander and Drucker, 2008, among others) and methods of assessing the opportunity costs associated with avoiding the use of crop genetic technology (Wale, 2008), as well as the opportunity costs of variety substitution (as per Zander et al, 2009 for breed substitution).

Institutional analysis in agro-biodiversity research is generally lacking (Smale, 2006). Contract and behavioural theory, institutional economics and transaction cost economics are also potentially promising and untapped areas to better link up economic analysis to genetic resources policy design. These areas of economics are becoming increasingly important as the management of genetic resources increasingly depends on crucial institutional and organizational decisions. When involving farmers as part of *in situ* conservation activities, principal-agent problems might arise (due to the existence of asymmetric information) and agent-based models could be helpful to assess farmers' willingness to deliver conservation services.

Given that most developing countries are at a fairly early stage in the formulation and implementation of genetic resources policy, the effectiveness and impacts of policy initiatives and conservation strategies on crop diversity remains to be assessed. The policy landscape itself may also undergo significant changes over time. Potential policy impact pathways have to be clearly established to better mainstream genetic resources policy-making and permit economic analysis to become more prominent in this process. Thus, another area for the future is crop diversity impact assessment and evaluation of genetic resource policies and conservation strategies.

Effective policies to stem the loss of traditional varieties of crops will require improved tools and the capacity to properly account both for the values associated with the services and the benefits derived from agro-biodiversity, as well as to design appropriate instruments to capture such values. There is a need to further improve economic methodologies and develop decision-support tools, combining economic concepts/data with ecological or genetic concepts/data. Without such tools, cost-effective interventions can be neither designed nor implemented.

References

Dasgupta, S., Hamilton, K., Pagiola, S. and Wheeler, D. (2008) 'Environmental economics at the World Bank', *Review of Environmental Economics and Policy*, vol 2, pp4–25

Gruère, G., Nagarajan, L. and King, O. (2009) 'The role of collective action in the marketing of underutilized plant species: lessons from a case study on minor millets in South India', *Food Policy*, vol 34, no 1, pp39-45

Mayrand, K. and Paquin, M. (2004) 'Payments for environmental services: a survey and assessment of current schemes', Report elaborated by UNISFERA for the Commission for Environmental Cooperation of North America, Montreal

Smale, M. (ed.) (2006) *Valuing Crop Biodiversity: On-farm Genetic Resources and Economic Change*, CABI Publishing, Wallingford, UK

Wale, E. (2006) 'What do farmers financially lose if they fail to use improved seeds? Some econometric results for wheat and implications for agricultural extension policy in Ethiopia', *Ethiopian Journal of Economics*, vol 12, no 2, pp59–79

Wale, E. (2008) 'A study on financial opportunity costs of growing local varieties of sorghum in Ethiopia: implications for on-farm conservation policy', *Ecological Economics*, vol 64, no 3, pp603–610

Wunder, S. (2008) 'Payments for environmental services and the poor: concepts and preliminary evidence', *Environment and Development Economics*, vol 13, no 3, pp279–297

Zander, K. K. and Drucker, A. G. (2008) 'Conserving what's important: using choice model scenarios to value local cattle breeds in East Africa', *Ecological Economics*, no 68, pp34–45

Zander, K. K., Drucker, A. G. and Holm-Müller, K. (2009) 'Costing the conservation of animal genetic resources: the case of Borana cattle in Ethiopia and Kenya', *Journal of Arid Environments*, vol 73, nos 4–5, pp550–556

Index